I0818177

Beekeeping Success

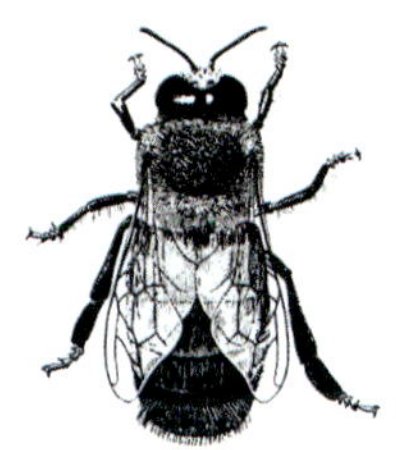

Robert Owen, founder of Bob's Beekeeping Supplies, has a PhD in honey bee diseases and has educated thousands of beekeepers across Australia, New Zealand, Canada and the US, as well as teaching beekeeping and advising governments on queen rearing, honey extraction, pollination and managing pests and diseases in Nepal, Bangladesh, Nigeria, Benin, Malawi, Ethiopia and Lebanon. A sought-after speaker and mentor, Robert blends science with hands-on expertise to help hobbyists succeed.

Jean-Pierre Scheerlinck is a researcher, educator and beekeeper whose scientific insights into hive health and bee behaviour make this book both rigorous and accessible. Together, they have drawn from decades of combined experience to write a reliable, down-to-earth guide that addresses real-life beekeeping concerns.

Beekeeping Success

START RIGHT
GROW STRONG COLONIES
HARVEST MORE HONEY

Robert Owen

Jean-Pierre Y. Scheerlinck

First published 2026

Exisle Publishing Pty Ltd
PO Box 864, Chatswood, NSW 2057, Australia
C/o Shortland Chartered Accountants Ltd, Level 9, 51 Shortland Street, Auckland 1010, New Zealand
www.exislepublishing.com

A CiP record for this book is available from the National Library of Australia.

ISBN 978-1-923011-28-1

Designed by Mark Thacker
Typeset in Minion Pro regular 10.75 on 14.5
Printed in China

This book uses paper sourced under ISO 14001 guidelines from well-managed forests and other controlled sources.

10 9 8 7 6 5 4 3 2 1

Contents

Contents *continued*

Contents *continued*

Contents *continued*

INTRODUCTION

Beekeeping is a popular and rewarding hobby practised by people from all walks of life and in many countries. Most beekeeping books devote chapters to the essential tasks, and finding relevant information for a specific problem or activity may be challenging. By laying out the book to address particular issues or questions a beekeeper may have, we worked to provide a reference that is both convenient to use and answers many common questions without having to trawl through many chapters looking for relevant answers to your questions.

Many new beekeepers feel daunted by their new hobby, particularly if they do not belong to a beekeeping club or do not have friends who are experienced beekeepers. By formatting the book to answer specific questions, new and experienced beekeepers can realise that beekeeping is not difficult and that most problems have easy-to-implement solutions, thus making their hobby more relaxing and rewarding.

The book is structured as a series of questions and answers. Because of this, the answer to some questions may be similar, but not identical, to the answer to another question. We have worked to eliminate duplication, and each answer should uniquely answer the relevant question.

In this book, we have used the terms colony and hive interchangeably. This is not technically correct since the colony is a family of bees, while a hive is a box where they live. In typical beekeepers' usage, however, hive and colony are used interchangeably, and we have followed this convention here. Also, a colony of bees can either be wild (or feral) or managed by a beekeeper; feral colonies live in nests, while managed colonies live in hives.

If any errors are noticed or you have suggestions for improvement, please send them to:

Robert Owen
beediseases@gmail.com

Robert Owen
Jean-Pierre Scheerlinck

CHAPTER 1

Getting Started

A beekeeper at peace with his hives.

Technically, a hive is a box where beekeepers keep a managed colony of bees, while a nest is where a feral colony lives. A colony is a collection of bees living in a hive or a nest. Many beekeepers use the terms 'hive' and 'colony' interchangeably, although this is incorrect. For convenience, however, in this book, we use the term 'hive' to mean 'colony' since this is standard usage amongst beekeepers.

What is the best time to start beekeeping?

Problem: Bee colonies can be acquired anytime, but in which season is it best to start beekeeping?

Solution

The best season to acquire bee hives depends on the climate in your area. If your region has a spring where many flowering plants bloom, this would be an excellent time to start beekeeping. Later in the year, it would also be good if flowering plants were abundant. A long late spring and summer with many consistent sources of pollen and nectar is optimal. This gives your new colony time to strengthen and prepare for winter. If you try to establish your new colony near a winter shortage of food sources, you may need help keeping your colony alive. Similarly, in geographical areas where you have hot, dry summers with few flowers, ensure that the hives you purchase in spring have enough supplies to last through the flowerless summer. The seasonality of flowering plants tends to be reduced in cities because keen gardeners often grow a diversity of plants that flower during an extended season (frequently only having no flowers around them when snow is present, which may never be the case in many parts of the world).

Buying an established colony with a local beekeeper or club support is the safest option. With this strategy, it is possible to start beekeeping during less bee-friendly times of the year. However, there are three distinct

A hobby beekeeper inspecting his hive.

advantages of starting in spring: (i) it is more likely that beekeepers will have surplus hives due to the need to control swarming, possibly by splitting hives; (ii) the beginner beekeeper will generally like to start with a smaller hive or well-established split because it is easier to transport and is less daunting to manage, (iii) depending on your area, the food resources are more likely to be abundant, (iv) a well-established split with a new queen is very unlikely to swarm until the following spring, giving you more time to learn swarm management techniques (see section on 'Swarming'). Installing packages without local support is the riskiest strategy (see 'How do I successfully install a package of bees into a hive?'). It is best attempted early in the honey-producing season, giving the colony time to build up numbers and store honey and pollen before winter.

How should I start as a beekeeper?

Problem: I want to become a beekeeper but must figure out how to start.

Solution

Many potential beekeepers have little, if any, prior knowledge or experience with beekeeping. Perhaps by talking with friends or workmates who are already beekeepers or watching documentaries about the problems honey bees face, many new beekeepers wonder how they will learn to keep these fascinating creatures and harvest delicious honey every year.

Many new beekeepers are fortunate to have friends who are also beekeepers and can gain knowledge and practical experience from them. Most experienced beekeepers take pride in their hobby and are usually willing to share their expertise with other beekeepers. Alternatively, particularly in larger towns, many beekeeping clubs are available that cater to both new and experienced beekeepers. They often have a dedicated section for new beekeepers, providing beginners' courses and hands-on experience. We recommend that anyone wanting to start keeping bees should contact a local beekeeper or club and experience first-hand what it feels like to handle bees by donning a bee suit and helping with a hive inspection. That hands-on experience will help determine whether beekeeping is something for you. Many beekeeping clubs are on the web if no club is near you. While not offering hands-on experience, they can attract members from across the country, even from other countries, and provide a more comprehensive range of experiences than is often available from local clubs.

Honey bees are hardy creatures, and it is difficult to disturb their daily lives. For many potential beekeepers, buying an established colony in a hive and locating it on their property is a good start. Selecting a suitable location for you, your neighbours, and the bees may take some thought (see 'Where can I place my hive?'). You will also have to quickly learn about minimizing swarming (see chapter on 'Swarming'). Advice from a local club, beekeeper, or the person from whom the bees were bought, can help. While relocating hives within the same property is possible, it is not trivial and requires great care (see 'How can I move my hive to a different location?'). So, choose your initial location wisely.

Is beekeeping expensive?

Problem: I have limited funds and would like to recover costs and build my hobby with honey production and sales proceeds.

Solution

The setup cost for a new 10-frame hive, plus protective clothing and tools, would be between US$240 and US$430, depending on whether you buy just a brood box or a brood box and a super. These figures are from Dadant (U.S. costs). Bees are not included if all the equipment is purchased as a set. These figures were accurate for November 2025. A typical package of bees costs between US$130 and US$200. This gives a total of between US$370 and US$630. These figures do not include extracting equipment, which can run from US$30 to hundreds of dollars, depending on whether you extract by hand or buy an extractor. There may be other setup costs, such as syrup or pollen if the colony needs feeding.

The ongoing costs may be up to US$100 a year, including medication and pest management supplies.

If you join a bee club or search the internet, you may be fortunate to buy used equipment and bees from a beekeeper who is giving up the hobby or downsizing. Be careful when buying used equipment; it should be free from pests and pathogens. The best way to ensure that equipment is sterile is to irradiate it before use. Irradiation is expensive and can only be performed by a commercial irradiating company, which may not be available in your area.

Revenue from the sale of honey varies widely between locations and years. The average beekeeper may expect to harvest between 35 kg and 70 kg (70 lb to 150 lb) per hive yearly. Depending on how you sell your honey and its quality, you can expect to earn between US$4 and US$12 per lb, giving a yearly gross revenue of between US$280 and US$1,800 from a single hive (in 2025). Some beekeepers also sell bee packages, queens, or nucs (a nuc is a small hive used to raise a queen. See 'Where can I obtain bees?' and 'How do I successfully install a package of bees into a hive?'), but that comes at the expense of honey production. Indeed, it is possible to produce honey or bees, but not both simultaneously.

Where can I obtain bees?

Problem: **I want to keep bees but do not know where to buy my first colony.**

Solution

The first place the authors would ask would be either at a local beekeeping supply store or at a local beekeeping club. Depending on how bees are sold in your area, either would likely have colonies for sale or know a local beekeeper selling hives with colonies or packages. Both groups are likely only to sell better colonies that are free of diseases. Another source would be a local swarm catcher. Again, a local beekeeping supply store, beekeeping club, or even local police or government agencies may know of a swarm catcher. An internet search may also provide details of a swarm collector. Although swarms are a good source of bee colonies since swarms are collected and then sold after a short period of time, swarm catchers often do not have time to check for disease or that the queen is a good egg layer with gentle characteristics (see 'What should I do if my bees are aggressive?'). However, more reputable colony providers may sell a swarm re-queened with a good quality queen.

Future beekeepers who live near larger towns or in agricultural areas where many beekeepers reside usually have little difficulty obtaining their first colony. In the US, future beekeepers who live in more remote areas may be able to order a package or colony and have it delivered by courier. Although in many countries, particularly in North America, packages of bees delivered to a remote address is the norm, the difficulty with this for a future beekeeper is to transfer the colony to a hive without harming the colony or the queen (see 'How do I successfully install a package of bees into a hive?'). The assistance of a local, experienced beekeeper would greatly aid this transfer and help ensure the successful establishment of a robust and healthy colony. To give packaged bees their best chance to get established, obtaining packaged bees in spring or early summer is preferable, allowing them plenty of time to expand, build comb and produce stores for winter.

In any areas with either managed or feral colonies of bees, there will

An urban beekeeper inspecting his hive.

be swarms every spring and often during the summer. With the aid of an experienced beekeeper, it is often easy to collect a swarm and rehouse it into a hive. Also, if time is not a concern for you, an easy option is to place an empty hive on your property with swarm attractant, or lemongrass, in the hive. Scout bees from a swarm seeking a new home may detect the attractant, check out the empty hive and get the rest of the swarm to move in (see 'How do I catch a swarm using a bait hive?'). For this to be successful, a frame containing undrawn foundation or drawn comb, preferably brood comb, must be placed in the hive for the bees to start building cells. If frames are not included in the hive, the swarm will build comb from the roof, resulting in a messy hive structure development and making it difficult to open the hive without destroying the nest.

How do I successfully install a package of bees into a hive?

Problem: I have purchased a package of bees, and I want to install it in a hive.

Solution

A package of bees is rehoused by first shaking the bees into a hive with frames set up with foundation. This is a daunting task for a beginner, and it is preferable to do it with the assistance of an experienced beekeeper. The queen cage is placed between the frames with the sugar-plugged hole facing upwards. The bees will eat through the sugar and release the queen. The sugar plug is pointed upwards so that if one of the helpers in the queen cage dies, she would not fall into the exit passage and block the queen's escape.

Adding a package of bees to an empty hive.

The packaged bees released in the hive need to be fed sugar water at a 1:1 ratio by weight every few days for at least two weeks to facilitate the building of the comb and the laying of eggs in the freshly built comb (see 'Should I provide syrup for my bees?').

Establishing a hive from a package of bees is much more complex than from a nuc and we would encourage beginners to purchase a nuc instead of a package of bees.

How do I successfully establish a bee colony from a nuc?

Problem: I have purchased a nuc of bees and want to establish them in a hive.

Solution

A nuc is effectively a tiny hive that contains four or five full-depth frames. As such, the hive can be transported to its final destination (see 'Where can I place my hive?'). Be careful to choose a good spot for the hive from the start, as moving an established hive is not trivial (see 'How can I move my hive to a different location?'). Once the hive is in place, ensure it is leaning slightly forward to allow water to flow out. You can use a spirit level or a smartphone with a level app.

When all is set, open the entrance and walk away. The bees will come out in relatively large numbers and circle the area to orient themselves. After a few hours, things will have settled down. In the next day or two, open the hive to check that there is fresh brood and that the bees have sufficient supplies. If supplies are low or if some frames need to be drawn out, consider feeding the bees sugar water at a ratio of 1:1 by weight.

Since the nuc is a small hive, you can continue to manage it until all frames are covered in bees. When you open the lid, the bees cover more than 50 per cent of the top of the frames. At that time, it is necessary to migrate the nuc into a full-size hive. Place the new full-size hive (bottom board, lid, and box) where the nuc was, moving the nuc one to two metres (approximately five feet) away. Prepare the new hive by opening it and placing empty frames with foundation along the sides. Smoke and open and transfer each frame carefully, keeping the same orientation and sequence of the frames. Close the lid and leave the hive alone for at least one week before resuming routine beekeeping inspections.

What are the local regulations about keeping bees?

Problem: My community may restrict my beekeeping activities. I need to find out what regulations apply.

Solution

Most beekeepers are fortunate that their communities tolerate beekeepers and sometimes even encourage keeping hives in the area. There may be restrictions on the number of hives you can keep in any given location, so check with your local government office, agriculture department, beekeeping club or beekeeping supplies store for advice. The most common restriction is that your colonies do not interfere with other people in the community, such as placing the hives next to footpaths so that passers-by could get stung, keeping gentle strains of bees and minimizing swarming. A regulation that is likely to apply to all beekeepers in your area is to monitor your colonies' health and treat any diseased colonies. It would be best, and sometimes a legal requirement, to keep bee strains that are not aggressive and re-queen hives that show aggressive tendencies. Regulations such as these can be obtained from an on-line search or by contacting your local agriculture department. There may be extension officers or agriculture inspectors who will be knowledgeable about the rules that apply to beekeepers. In our experience, government apiary inspectors are helpful and work to ensure that hobby beekeepers are informed about any local regulations. If there is a dispute between a beekeeper and a neighbour, so long as you comply with local laws, an apiary inspector will generally work to smooth the situation to your advantage. If you are not complying with local laws, the inspector is likely to come down on the side of your neighbour, often insisting you make changes to your apiary. Also, in many countries, it is compulsory to register as a beekeeper so that authorities can contact you if required, for example, to warn about disease spread or the use of pesticides in your area.

What should I do if my neighbours complain about my bees?

Problem: My neighbours complain that my bees are a nuisance, even though they are not disturbing them. I need advice on how to overcome their concerns.

Solution

A government apiary inspector told me that while only about 5 per cent of the population are allergic to bee stings, when neighbours discovered a hive was on their street, about 80 per cent claimed to suffer severe allergic reactions. For this reason, many urban beekeepers prefer to keep their hive hidden from view, believing that discretion is the better part of valour. If you follow local regulations for keeping bees, any complaint to the council will likely be dismissed. If you are not following local rules, you can expect local authorities to be more sympathetic to your neighbour.

Proactive diplomacy, though, can often calm a neighbour's concerns. Some beekeepers first place empty hives for a month or two as 'placeholders' where they plan to have hives. The people who tend to be overeager to lodge complaints immediately fire off their complaints, hereby discrediting themselves as there are no actual bees there yet. They are left with little to say once bees do arrive.

There are various steps that other beekeepers have used successfully to defuse this type of situation:

Swarms often land in places that are difficult to access and may become a nuisance to neighbours.

- Talk with your neighbour in a friendly manner about their concerns and listen without interrupting.
- Provide counterarguments and explain that bees, unlike wasps, rarely disturb people. Explain the difference between Africanized bees and European bees, that Africanized bees are localized in South America and parts of the US and that when you live in an area where Africanized bees occur, your bees must be screened not to contain any of the aggressive genetics.
- If the neighbour is a gardener, explain that bees are crucial as pollinators, and with them, most plants in their garden will have a more prolific fruit set.
- Be a good neighbour. Offer to give them some honey every year. Garden honey tastes nicer than shop honey, and they may come back more regularly, allowing you to sell some of your local produce.
- Keep a spare veil or jacket and invite them to inspect the colony with you. An educational visit often alleviates any concerns they may have.
- If the colony shows signs of aggression, requeen with a gentler strain of bees without delay.
- Ensure plenty of water on your property for the bees to drink. If foragers discover a nearby swimming pool is the only water source, it will be difficult to wean them off it.
- Keep your colonies hidden from view behind shrubs, a fence or a garden shed. Remember the old saying, out of sight is out of mind. You can also consider painting your hives green, to avoid detection.
- Manage your colony so that it does not swarm. A swarm resting on your neighbours' clotheslines is not favourable to public relations.

CHAPTER 2

Equipment

Basic tools, bee hives, boxes and glove.

A good bee suit makes beekeeping enjoyable.

What equipment do I need to start?

Problem: **What equipment should I buy as a beginner beekeeper?**

Solution

It is unnecessary to buy lots of equipment since much of the equipment for sale by suppliers is not essential, especially for a beginner. A list of the equipment that would be needed to start is:

The hive: Your hive should allow inspection of the comb, which is generally achieved using removable frames or top bars. Hives can be made from various

Assembling a hive.

materials and come in various sizes (see 'What type of hive should I buy?').

Protective clothing: Good-quality protective clothing includes a bee suit or veil, loose trousers, gloves, extra thick socks and sturdy boots.

Tools: The kit includes a hive tool to remove frames or top bars from the hive and a smoker to calm the bees.

Extracting equipment: While not essential, this dramatically facilitates honey collection. Alternatively, a Flow Hive can be used (see section on 'Flow Hives').

Equipment needed to start beekeeping. From left to right: bee brush, queen catcher, scratcher, plunger to mark queen, American design hive tool, smoker, extracting knife, J-tool and grip to hold frames.

What are the components of a typical hive?

Problem: A hive is composed of distinct components that are used for different purposes.

Solution

A hive consists of several components designed to house and support a colony of honey bees. The exact design and materials can vary, particularly between a Langstroth hive and other designs such as top bar, Warre or Flow Hives, but here are the essential components commonly found in a Langstroth hive:

Stand: Although many beekeepers prefer to leave their hives resting on the ground, others raise them about 0.2 m to 0.4 m (8 to 16 inches) above ground. This practice has several advantages. The entrance to the colony is less likely to become blocked with long grass; bees naturally make their nests above ground, although usually many metres above ground; raising the hive makes inspections easier; and, if the hive is higher, removing the super during extractions is easier. A hive stand can be constructed in many ways, from simply placing the hive on bricks to more elaborate purpose-built hive stands.

Bottom board or base: This is the hive base, providing an entrance for the bees and serving as a platform on which the hive rests. There are many designs for the base, and it is the part of the hive that will vary the most, depending on the beekeeper's choice and its role. Some designs only function as a solid base, while others include additional ventilation or traps for *Varroa* mites or small hive beetles.

Brood box, or hive bodies: These are the boxes where the queen bee lays eggs and the worker bees raise brood (larvae and pupae). There may be one or two brood boxes in a hive, depending on the strength of the colony, and they contain frames for bees to build comb in which eggs are laid and larvae reared. Brood boxes usually contain either ten frames or eight frames and come in different depths (see 'What are the different types and sizes of hive boxes?'). If a brood box contains eight or ten frames, the super above must also be the same size, but not necessarily the same depth.

Queen excluder: This optional component is a metal or plastic grid that prevents the queen bee from entering the honey supers, allowing workers to go through it. This ensures the queen cannot lay eggs in the honey supers. Our recommendation, particularly for a beginning beekeeper, is to use an excluder as honey extractions are easier, and you are less likely to kill the queen during inspections or extractions. It should be noted that drones (male bees), larger than female workers, can also not pass through the queen excluder. Thus, any drone brood placed on top of the excluder will result in the emerging drones being trapped above the excluder, unable to leave the hive. This should be avoided as, over time, it results in a layer of dead drones on top of the queen excluder.

Super (Superstructure): These are boxes placed above the brood boxes where bees store surplus honey. Depending on the amount of honey the colony produces, say during a spring honey flow, the hive may include more than one super. Supers also contain frames for comb building. They usually contain either eight or ten frames and come in different depths. While boxes of different depths can be mixed within the same hive, eight- and ten-frame boxes should not be mixed within the same hive.

Inner cover or hive mat: A hive mat, or in North America, an inner cover, is a mat or flat board that rests between the uppermost hive body and the lid. It provides insulation and may provide additional ventilation for the hive. The inner cover often has an opening in the middle to allow bees to access the space between the inner cover and the lid, preventing pests, such as ants and small hive beetles, from establishing themselves there. An essential use of an inner cover is to stop the colony building comb under the lid. In much of the world, beekeepers place a mat of sackcloth, linoleum or other flexible material under the lid instead of an inner cover. The dimensions of the hive mat may be slightly smaller than the inside dimensions of the super, which provides additional ventilation and allows the bees to monitor the area between the mat and the lid for pests such as small hive beetle.

Lid or outer cover: The lid or outer cover is the topmost part of the hive,

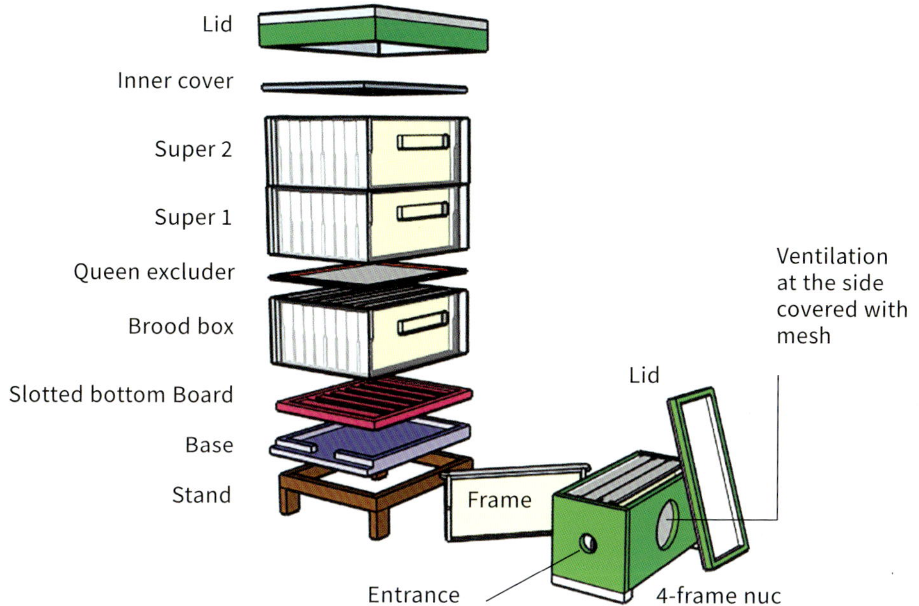

Parts of a Langstroth hive.

protecting it from the weather and other elements. The lid also comes in several designs and may extend beyond the edges of the super to provide extra protection against rain (a telescopic lid) or may be flush with the super (a migratory lid). Some lids are pitched, which enhances the aesthetic appeal of the hive. However, many beekeepers prefer flat lids as they are simpler/cheaper and allow for the turned-up lid to serve as a place to stack the hive boxes when the hive is opened. Depending on the climatic zone where the hive is kept, some lids have inbuilt insulation.

Frames: Inside a Langstroth hive are frames on which the colony builds comb to rear larvae or store honey and bee bread (made from pollen). The outside edges of frames may be made of wood or plastic and support the comb. Beekeepers usually insert foundation inside the frame to assist the colony in making a comb. The foundation may be made of beeswax or plastic and is embossed with a honeycomb pattern the size of female brood cells. Bees do not need a foundation inside the frame, but its inclusion speeds up the build of the comb and produces a consistent, orderly layout that assists the beekeeper in removing frames during inspections and honey extraction. Wires are often used between the sides of the frame to strengthen the comb and facilitate spinning of the frames for honey collection. These wires do not interfere with comb building; bees build their comb around them, whether foundation is used or not. In many countries, the use of frames or removable top bars is compulsory to allow for the inspection of the brood for bee diseases. Hence, traditional skep hives (typically resembling an upturned basket made of straw) are banned in these areas. Frames also greatly facilitate the collection of honey without destroying the colony. The space between consecutive frames is essential to ensure that the bees are building parallel combs and is called the bee space (typically 5-9 mm or 1⁄4–3⁄8 inches). The discovery of the bee space (i.e., the precise spacing of the comb in a hive) by Langstroth was undoubtedly one of the major developments in beekeeping, opening the way to the Langstroth hive with removable, widely used frames.

What type of hive should I use?

Problem: I need to choose between the different available hive designs.

Solution

We recommend starting with an eight- or ten-frame Langstroth hive with a brood box and a super since these hives are widely available. Also, the Langstroth hive is the most popular design amongst hobby and commercial beekeepers in most of the world. Since most beekeepers use this design, advice from other beekeepers, clubs and online sources will be readily available and applicable to your apiary. Since supers full of honey can be heavy, you need to decide on the depth of the super. Full-size ten frame deep supers can weigh upwards of ~20-25 kg (~44-55 lb), which may be

Popular hive designs include Langstroth, top bar, and Flow Hive.

too heavy for many people to lift. An alternative is to use half-size or ¾-size (called 'ideals' in the UK and Australia, and 'Shallows' in the US) supers, which will be lighter (see 'What are the different types and sizes of hive boxes?'). The disadvantage of using supers of different sizes for the brood box is that frames will not be interchangeable. Alternatively, instead of a full-size brood box, you could use two or more smaller brood boxes, the same size as your supers, making both frames and hives interchangeable.

Many other designs can be considered. These include the Flow Hive, top bar and Warre hives. The Flow Hive is gaining in popularity, and its unique design refers only to the super, not the brood box. Opinions on the practicality of the Flow Hive vary. There are strong supporters and strong critics, each with valid arguments. See the Flow Hive section later in this book for more details.

Natural beekeeping is also popular, using a top bar hive, sometimes a Warre hive, or even the more exotic sun hive. However, equipment is less readily available, and your woodworking skills would be useful if you decided to use either of these designs (see section on 'Natural beekeeping'). In any case, it is often a legal requirement that comb be able to be removed for inspection.

Unless you have a good mentor or another source of natural beekeeping support, we recommend starting with a Langstroth hive since equipment and support are more readily available. Later, as you gain experience and confidence, you can include other hive designs.

How many brood boxes and supers should I use?

Problem: I wonder how big the hive needs to be for the different seasons and how many hive boxes I need to buy.

Solution

The number of boxes that are used depends on the size of the boxes, both in terms of the number of frames per box (ten frame hives are used in most of the world, while some regions use twelve frame or, less frequently, eight frame hives) and the depth of the boxes (see 'What are the different types and sizes of hive boxes?'). We will use a standard deep box size to have a useful conversation on the topic. It is easy enough to convert that when using different sizes of boxes. For example, if using shallows, we can approximate two shallows for one deep box.

The size of the hive will vary with the seasons and when beekeepers harvest honey. Other critical factors are the length of winters and climatic conditions. Indeed, during harsh winters, bees rely on stored honey for survival. In other areas with hot and dry summers, the bees may need to store honey to survive this food-less, dry season.

Nevertheless, general rules of thumb are a good starting point for discussions with local beekeepers. One general rule is that the hive needs to be as small as possible to accommodate the number of bees present and anticipated to be present in the coming weeks, as well as the required honey stores for the bees to survive until nectar and pollen are again available. Too much space and the bees will spend excessive energy keeping it at a suitable temperature, exhausting their stores and starving. If the space is too small, there is a risk that the hive will swarm in spring and early summer. Indeed, crowding is a clear signal that the hive needs to reproduce and may soon swarm. To start winter, one should have a two (deep) box hive for temperate climates. One box with brood at the bottom and some reserves and (preferably) a full box of honey at the top. If it is becoming clear that this is unlikely to be the case, it is best to start feeding the bees sugar syrup in the autumn (see 'How do I provide syrup for my bees?'). Should there only be, say, five frames of honey, it is better when packing down in the autumn to cram the bees into one crowded brood box than into two half-empty boxes. See 'How do I prepare my hives for winter?' The chances of the hive surviving are much greater when a small colony does not have to keep a large hive warm. In some climates, it is possible (although not necessarily advisable) to overwinter small colonies in five frame nucs. If one must overwinter in a small hive, it is imperative to quickly expand it by adding a box as soon as the colony starts growing in spring. Indeed, as the number of bees increases rapidly once nectar and pollen are coming in, the crowded hive is very likely to want to swarm, even if the total number of bees in the hive is likely too small for this to be successful. Hence, overwintering in a small hive, while best for the bees, requires vigilance in spring.

Another rule of thumb is that if one opens a hive and there is instantly a layer of bees covering the top of the frames, it is time to add another super or, in the case of a nuc, to expand it into an eight- or ten-frame hive. Nevertheless, in that case, it is still a good practice to check that the bottom box is not largely empty. Indeed, especially when not using a queen excluder, the bees will tend to gather at the top of the hive, which is warmer when the weather is cold. If there are more than three substantially empty frames (i.e., no pollen, brood or nectar/honey) in the bottom box, it is best to delay adding a super.

In summer, just before harvesting, it is not unusual to have hives that are four or five deep boxes high, depending on the location and the nectar flow in the area at the time. In that case, there would likely be two deep brood boxes and two or three deep supers on top (most likely with a queen excluder between the brood boxes and the super in preparation for a harvest; see 'Should I use a queen excluder?'). Higher than that, the hive becomes unwieldy, and one should consider harvesting the honey in the boxes above the queen excluder before adding another super.

Should I use synthetic or wooden hives and frames?

Problem: I wonder what are the advantages and disadvantages of using plastic versus wood hives and frames.

Solution

Synthetic hives can be made with different materials, usually polypropylene or polystyrene (Styrofoam), and have the advantage of needing less maintenance. Polypropylene hives do not require painting, while polystyrene hives do. Plastic Langstroth hives are available in different designs,

A plastic hive has many advantages over a wooden hive.

including innovative and practical designs for the bottom board or base and hive cover or lid.

The authors prefer wooden hives since this is the material used by feral colonies and, thus, is more natural. Many beekeepers, however, only use plastic and would not switch back to wood. The decision is best left to the beekeeper since both materials have good and bad points. There is also no problem using plastic for the brood box, base, top cover, lid and wood for the super, or vice versa. Plastic Langstroth hives are designed to be used with a wooden brood box or super, and the parts are interchangeable. In general, synthetic hives have a better isolation coefficient (e.g., polystyrene has an R-factor of 6. In contrast, wood has an R-factor of 1.2), which may make a big difference if you keep your bees in areas that experience more extremely hot or cold climates.

Frames are also available in wood and plastic. Some plastic frames include only the sidebars, a top bar, and a bottom bar and do not include a foundation. Depending on their management philosophy, beekeepers can decide whether to use wax or plastic foundation. See 'Should I use foundation, and should it be plastic or beeswax?'

Should I use foundation, and should it be plastic or beeswax?

Problem: **Some beekeepers do not use foundation, while others use plastic or beeswax foundation. I would like to know the advantages and disadvantages of these different methods.**

Solution

The bees generally need to have somewhere to start their comb-building activities. Some beekeepers place a small foundation strip (say 2.5 cm (1 inch)) in the slot at the top of the frame, allowing the bees to do the rest.

In some cases, beekeepers even use a small wooden strip wedged into the groove at the top of the frame as a starting material. The advantage of this method is that it is cheap, and the comb the bees build is the comb they need at the time, making it more natural. This often results in more drone brood than the beekeeper may want. Indeed, drones can be considered a drain on the hives' resources since they do not produce honey and have no function except to fertilize a queen from another hive. Another drawback is that the comb built is often not as straight as the one built on foundation. In extreme cases, it is possible that the bees would build a comb across several frames, making it impossible to remove the frames without destroying part of the comb. To prevent this, always make sure that a new frame without a full foundation is placed between two fully drawn-out frames (i.e. a frame where the bees have made honeycomb to its full thickness) to guide the bees in making a straight comb within the confines of the frame.

If the beekeeper decides to use foundation, the question remains whether to use wax or plastic foundation. When using a plastic foundation, it is essential to encourage the bees to make a comb by coating the plastic with a thin layer of wax. This can be done using a roller or brush and applying molten wax to the plastic foundation. Be extra careful when melting wax that it does not overheat or is exposed to a flame as molten wax can be highly flammable and can ignite when overheated even if no flame is present (see 'How do I produce wax and what can I do with it?'). One way to achieve this is to melt the wax in a pot placed inside another pot with water in between. The temperature at which water boils, (100 °C or 212 °F) is below the flashpoint of the wax (204 °C or 400 °F). The main advantage of using a plastic foundation is that it is very durable and can be reused when the comb turns dark (see 'What should I do when the comb turns hard and dark?'). Just scrape off the comb, wash the plastic foundation with hot water, dry it and recoat it with a new layer of molten wax. Thus, while the initial cost is higher than the alternatives, in the long run, it is pretty cheap. The disadvantage is that given a choice, bees will generally prefer wax foundation to plastic foundation to build comb. Hence, it is often necessary to put frames with plastic foundations in the middle of the hive

Above: Plastic foundation inserted into a wooden frame.

Above right: A solid plastic frame.

Right: The Beeswax foundation is in a wood frame showing a stainless steel supporting wire.

to encourage the bees to use them and draw out the comb. Once the comb is drawn out, the frame can be positioned back where it best suits the beekeeper.

Wax foundation is more fragile and needs to be stored flat in a dry, cool place. If stored too long, it can dry out, becoming brittle, and this may require soaking in warm (not too hot to prevent melting) water. The advantage of wax foundation is that the bees like this foundation best and will quickly use it when requiring extra space. Wax foundation can't be recycled, so when the comb is old and needs to be replaced (say more than two to three years old, depending on how much it was used for brood) the entire comb will be cut out, and a new sheet of foundation used (see 'What should I do when the honeycomb turns hard and dark?'). This makes using wax foundation quite expensive in the long run. Beekeepers must be aware that wax can accumulate agricultural chemicals from the hive's environment and miticides that might have been used inside the hive previously. As such, the authors would encourage beekeepers to pay attention to the origin of the wax foundation they purchase.

We recommend that beekeepers experiment with different types of foundation (or just starter strips) and select the preferred method given their local conditions and level of expertise.

Should I use a queen excluder?

Problem: **There are advantages and disadvantages of placing a queen excluder between the brood box and the super. I am concerned that the excluder will hamper the free movement of bees through the hive and result in a weaker colony.**

Solution

Queen excluders are placed between the brood box and the honey super to reduce the likelihood that the queen will leave the brood box and move up into the supers to lay eggs in the comb. In our experience, about 90 per cent of beekeepers use queen excluders, although many good arguments exist for and against their use in a hive, as explained below.

Advantages:

- Using an excluder keeps the queen below the honey supers so that frames of honey can be removed without any concern about the brood being present.
- Since the queen is prevented from laying eggs in the super, the comb will remain white or light yellow for longer, without the unsightly dark wax caused by larvae skins, excreta and cocoons.
- Keeping the queen in the brood box, a restricted area of the hive, makes her easier to find when needed.
- Many workers will stay with the brood in the bottom brood box. This is an advantage when removing frames of honey since bees are more easily removed from honey frames than from frames containing brood.
- Pollen is usually stored next to the brood, so there will be little pollen that will discolour the honey in the super.

- Some procedures, such as the Demaree method of swarm control, require a queen excluder (see 'What is the Demaree method of swarm control?').

Disadvantages:

- Reduced airflow through the hive can cause overheating on scorching days. Brood cannot survive heat as easily as they can cold.
- Workers may have trouble getting through the excluder's grille and cannot store some of their honey in the super.
- Excluders are expensive and easily damaged.
- Some, mainly plastic, excluders with sharp edges around each mesh opening may damage workers' wings as they struggle through the grille.
- If the queen inadvertently gets through the excluder into the super and lays eggs, larger drones may get trapped above the excluder in the super when they emerge. These will die and form a layer of dead drones on top of the excluder, which can hamper the workers' movement through the grill.
- Worker bees come in all sizes, and a single gap width on a queen excluder may not be suitable for all the hives in your apiary. Queens come in different sizes, and the tiny virgin or newly mated queens may quickly get through some of the excluders.
- Using an excluder is for the sole benefit of the beekeeper and not the bee.
- The main issue is that the excluder constricts the size of the brood nest and reduces the capacity for growth.
- By confining the brood to the lower box, the hive expends more energy, keeping that box warm as heat rises. In winter, a hive without an excluder typically moves the brood as close as possible to the lid where it is warmer.
- It may also encourage swarming.

A queen excluder on top of a brood box stops the queen from moving into the supers to lay eggs.

Excluders are generally made of galvanised/stainless steel or plastic, although some suppliers also offer a bamboo version. Any of these materials is suitable for use as an excluder. Some plastic excluders have sharp edges that should be avoided.

Some beekeepers do not use an excluder until about six weeks before honey extraction, allowing the colony full access to the hive until this period. In about week six or earlier, an excluder is placed below the boxes containing most of the honey, ensuring the queen is below the excluder for at least six weeks before harvesting. This allows sufficient time for the brood trapped above the excluder to emerge. If there is a strong nectar flow, the empty cells will be filled with honey. During most of the year, however, the bees are free to decide how to organize the hive and are not impeded by the presence of a queen excluder.

Queen excluders are also helpful when honey produced for human consumption needs to be separated from the honey used by the bees. This is often the case when the beekeeper uses chemical treatment for bee diseases such as *Varroa* mites or (when allowed) American or European Foulbrood (see sections on '*Varroa* mite' and 'Other pests and diseases').

What equipment do I need to perform advanced beekeeping tasks?

Problem: Now that I have the essential equipment, I would like to move into more advanced procedures and require more specialist equipment.

Solution

As you gain experience as a beekeeper (after about a year), you will have the knowledge and skills necessary to perform more advanced activities. These include:

Catching swarms: For this, you need either a nuc or a spare one-box hive that can be sealed but still allow bees to breathe. The hive must be securely closed when moving in your vehicle, so clips holding the base, lid or straps are essential. A bee brush will be required, together with pruning tools, to cut away small branches if the swarm has settled in a bush. A water spray will be to spray the swarm to make them more docile and encourage them to go inside the hive on a warm summer evening before closing and moving the hive. See Chapter 11, 'Swarming', for more details.

Rearing queens: Many new and experienced beekeepers are interested in rearing queens, and this will provide hours of exciting activity to your hobby. There are many ways to rear queens, some of which are discussed in Chapter 13, 'Queen rearing'. Basic equipment needed to rear queens include one or more nuc boxes (also called a nuc hive) or spare one-box hives, queen cages with sugar plugs to hold newly emerged virgin or mated queens, a marker pen, queen spike cage to hold the queen to the comb so she can be marked or a queen piston for marking. If you decide to practise grafting, see 'How do you graft larvae to rear queens?'

Extracting honey: Many new beekeepers postpone buying an extractor until they harvest enough capped frames or use rented equipment from clubs and beekeeping supply stores. The selection and use of extractors are discussed in Chapter 14, 'Harvesting Honey.'

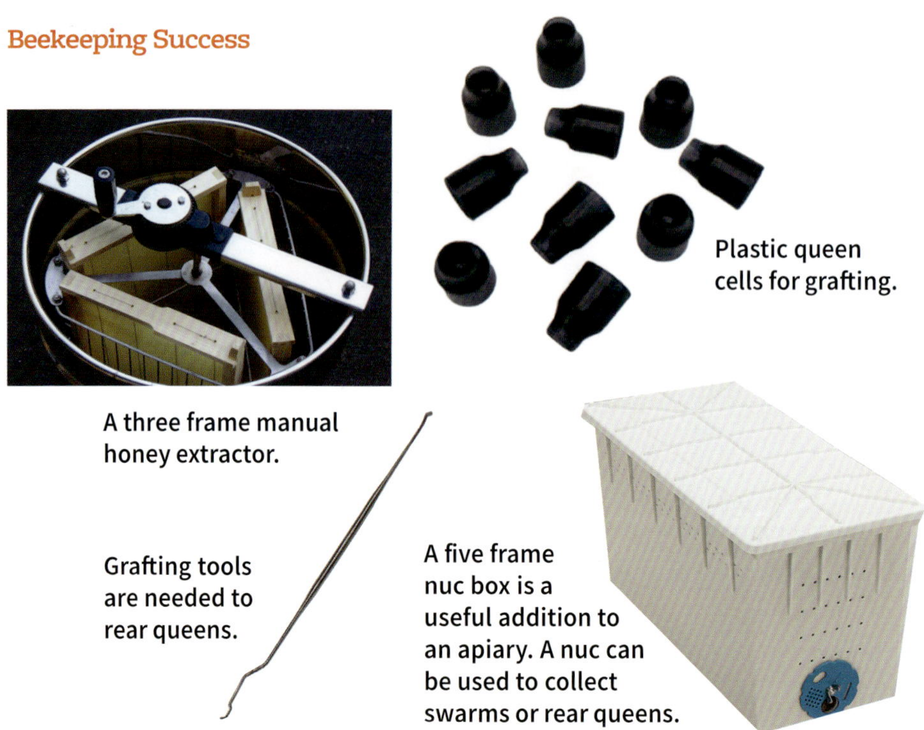

A three frame manual honey extractor.

Plastic queen cells for grafting.

Grafting tools are needed to rear queens.

A five frame nuc box is a useful addition to an apiary. A nuc can be used to collect swarms or rear queens.

Managing pests and diseases: Equipment needed to manage pests and diseases may not be optional but an essential part of basic beekeeping in your area. Almost certainly, you need to manage *Varroa* mites and possibly small hive beetles. For *Varroa*, perform a sugar shake or alcohol wash to monitor *Varroa* levels in your hive. Green drone frames may also be used to capture *Varroa*. See Chapter 18, '*Varroa* mite'. Tools needed to manage small hive beetles and other pathogens are discussed in Chapter 19, 'Other pests and diseases'.

Splitting and merging hives: Other hive manipulations include splitting hives or merging colonies, which are discussed in the sections 'How do I prevent swarming and produce more hives?' and 'I have two weak colonies; how can I merge them?'

Observing and showing bees: Beekeepers who regularly give talks to schools or clubs often keep an observation hive that usually contains a single frame of brood kept in a thin hive with Perspex sides. Participants can observe the bees and identify the queen, workers, drones, larvae and capped brood through the Perspex walls.

How do I prevent bees from building comb under the lid?

Problem: **My bees are building comb under the lid. This makes removing the lid difficult and causes broken comb to spread honey throughout the hive.**

Solution

Beekeepers insert either an inner cover or a hive mat between the top of the supers and the lid. This usually deters bees from moving into the space under the lid to store honey. See 'What are the components of a typical hive?' for details of hive mats and inner covers.

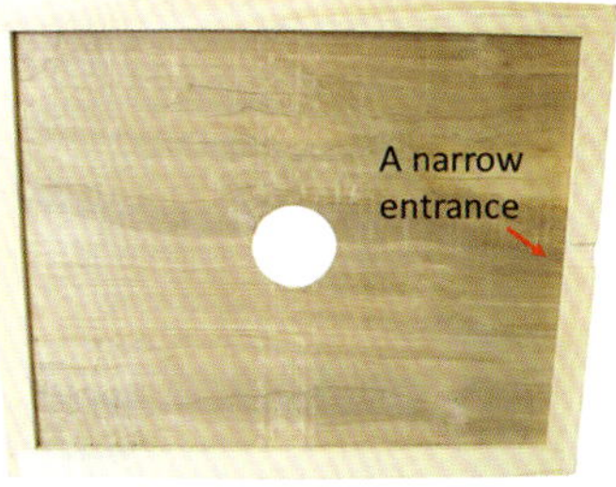

An inner cover placed under the lid prevents workers from storing honey in the lid. This top cover includes a large opening, which can also be used to merge and split colonies. The opening also allows bees to patrol the area above the top cover for pests such as small hive beetle (see 'What should I do when my hive is infested with small hive beetles?'). On the right edge of the top cover is a small entrance that foragers can use to enter or leave the hive. The entrance is only on one side of the inner cover, so it can be used as an entrance, or to seal the hive.

Hive mats can be made of lino, which is inexpensive and flexible.

What are the advantages and disadvantages of using a deep body?

Problem: **My colony is strong, and the queen lays more eggs than can be accommodated in a standard-sized brood box. Should I use two standard-sized brood boxes or buy a larger deep?**

Solution

Although the meaning of the names 'deep' and 'medium' varies, many beekeepers in the US refer to a deep as having a height of 24 cm (9 5/8 inches), while a medium-sized hive box has a height of 16.8 cm (6 5/8 inches); this provides the queen with more space in which to lay eggs (see 'What are the different types and sizes of hive boxes?'). Although using a larger, deep brood box may be convenient, the additional weight makes moving a deep difficult. Also, in our apiaries, we keep all brood boxes and supers the same size so that they are interchangeable. This is particularly important during spring manipulations, where, in our apiaries, the brood box is placed above the super, and the queen moves down into the newly located super (see Chapter 7, 'Spring and summer management') and 'What are the different types and sizes of hive boxes?'

What are the different types and sizes of hive boxes?

Problem: **There are many sizes and different designs of hive boxes available, and I want to know the difference between them.**

Solution

In nature, a swarm of honey bees will seek a site of approximately 50 litres (~13 gallons) to build a new hive. Many shapes and configurations are used

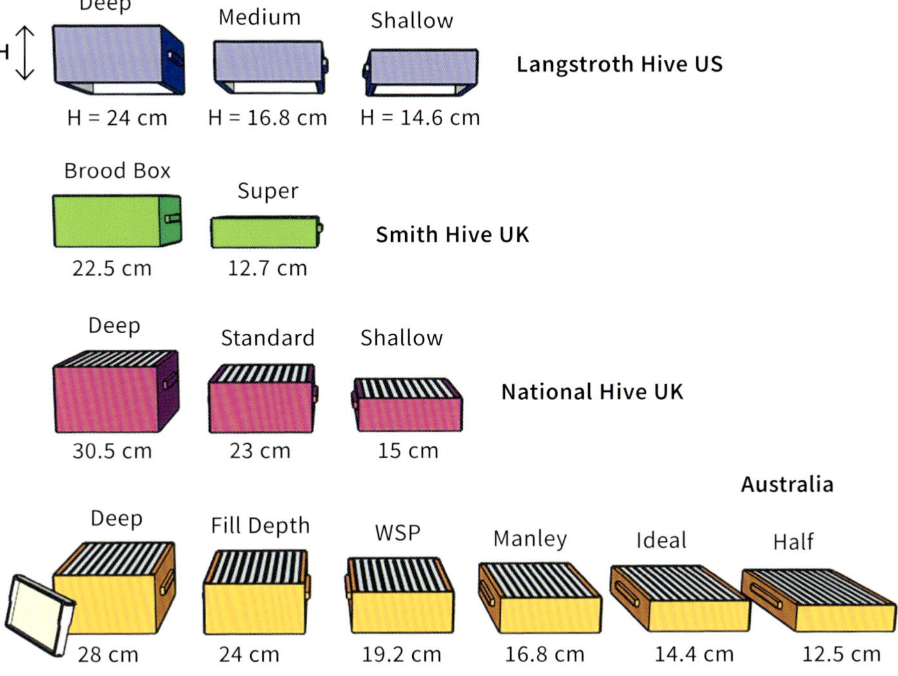

Frames matching each box size are generally 7.9–9.5 mm (5/16–3/8 inch), smaller than the depth of the hive box. In addition, there are also jumbo frames that fit the depth of two shallow boxes on top of each other (i.e., depending on the type of box used, the total height of combined boxes is 292 mm (11 1/2 inches)) and are about 11 inches (280 mm).

as the bees adapt their comb to the available shape of their new home. As beekeepers, we can use any configuration but prefer easy-to-make standardized hive boxes. Nevertheless, many different sizes are often linked to local historical standards. Hence, for ease of working with other local beekeepers (i.e., getting your first hive, information about hive management and availability of equipment such as centrifuges for harvesting the honey from frames), it is wise to consider what others are using in your area. Consult with a local bee club or bee equipment supplier for popular models. As a first step, it is always best to use these standardized hive boxes, which are often but not always Langstroth hives (see 'What type of hive should I use?'). Within the Langstroth hive design, there are eight or ten frame boxes (best to stick with either of these, but not mix them in your apiary), and for each, there are different depths (see 'Should the brood box be the same size as my supers?').

How do I install wires in my frames?

Problem: **The challenge when installing wires in a frame is to ensure the wire is tight.**

Solution

Tightening wires in a frame requires some tools to slightly bend the sides of the frame while the wires are stretched and attached. The sides are then released to allow the wood to regain its original shape, tightening the wire. While there are special wiring boards to perform this procedure, achieving the same result is possible using a quick grip bar clamp. Here is a step-by-step procedure:

1. Install brass eyelets into the holes in the side bars of the frames. You can do this using a needle punch or a nail.
2. Hammer two small nails partially into the frame to attach the wire.
3. Thread the wire (preferably stainless steel) through the brass eyelets from side to side across the frame.
4. Twist the end of the wire three to four times around one of the partially inserted nails and nail in completely in place to secure the wire.
5. Bend the end of the wires until it snaps off flush with the nail head.
6. Place the frame on the wiring board and clamp the side to bend slightly inwards. Alternatively, use a quick grip bar clamp to squeeze the sides of the frame together, slightly deforming the side throughout the process.
7. Use a pair of pliers to tighten the wire on each run of the frame. This will ensure that minor kinks do not form near each eyelet, helping with the overall tightness of the wire.
8. While stretching the wire, twist it three to four times around the second nail as tight as possible and hammer it until it is flush with the wood.
9. Bend the ends of the wires until they snap off flush with the nail head.

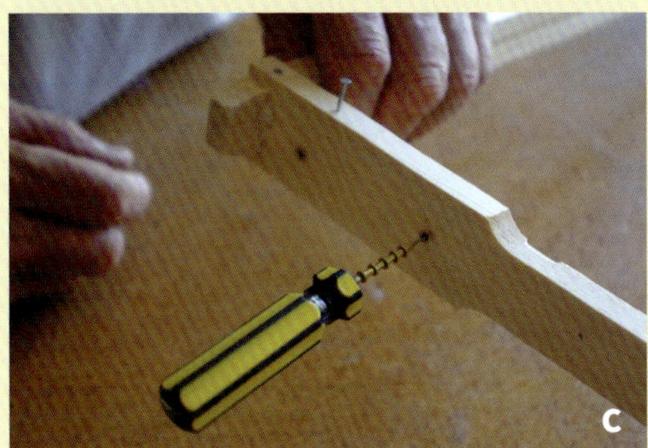

INSERTING WIRE INTO A FRAME

A Tools needed to make a frame.
B Assembling the wood parts of the frame.
C Insert eyelets into the frame's walls to prevent the stainless-steel wire from cutting into the wood and becoming slack. **D** A jig holds the frame and is ready to thread the wire. **E** A clamp bends the sides of the frame inwards; when released, the wire is taught.
F Wire is threaded through the eyelets.
G Tying off the wire on a tack or small nail.

10. Release the clamp or the tensioner on the wiring board. This will allow the sides of the frame to regain their original shape and tighten the wire in a fully stretched position. The wire should be as tight as a guitar string, so it keeps the comb in place during honey harvest using a centrifuge (see 'How do I harvest honey using an extractor?').

How do I install a wax foundation into a wired frame?

Problem: I want to install a sheet of wax foundation into a wired frame and embed the wires into the wax.

Solution

Most frames have wires across the frame to strengthen the comb and facilitate centrifugation of the frame for honey extraction. These wires need to be embedded into the wax of the foundation. This is best done using an electrical (12V) embedder, which heats the wire, allowing it to melt the wax while at the same time pushing down the wire into the foundation. A board is used under the frame, so the foundation and the wires are in the same plane.

An alternative to the electrical embedder is a wax foundation embedding wheel. This small tool has a wheel with a groove that can push the wire into the wax foundation. It works best when slightly warmed by dipping into hot water. Using this tool, a board and baker's paper under the wax foundation is essential.

While more expensive, the electrical embedder is more convenient, especially for larger apiaries, where many frames must be prepared.

If no foundation sheet is used but instead a small strip is placed at the top of the frame, the bees will build comb around the wire, effectively embedding the wire in the process (see 'Should I use foundation, and if so, should it be plastic or beeswax?').

EMBEDDING WAX FOUNDATION.

A The wire frame is ready to have the foundation attached. Resting next to a base. **B** Inserting foundation into frame. **C** Checking the foundation is in the groove in the top bar. **D** Placing the frame with foundation onto a base. **E** The foundation rests on the base, ensuring it is flat. **F** Manually attaching foundation to wire. **G** Attaching the foundation using an electric current to heat the wire between the first and last pins.

How should I light a smoker?

Problem: My smoker is difficult to light, and I want to know the best way to light it so that it keeps alight and produces plenty of smoke.

Solution

We advise always lighting a smoker, even if only a quick inspection is planned and the hive contains gentle bees. Indeed, a lit smoker becomes extremely useful when dealing with an unexpectedly aggressive hive, which could occur, for example, if a previously gentle hive has become queenless, resulting in more aggressive workers. Having to quickly light a smoker while dealing with aggressive bees and being stung should be avoided.

Our reliable method for lighting a smoker is highly effective. We use a barbecue lighter, holding it deep inside the smoker while puffing air through the fuel. We also place some crunched-up paper at the bottom of the smoker, below the fuel, as this quickly lights and spreads to the fuel. An alternative and quicker way is to use a propane torch purchased from a hardware store. Due to its heat, a propane torch quickly lights the fuel, and the flame can be directed through the fuel deep inside the smoker. A propane torch will quickly relight the fuel if the smoker goes out.

The fuel used is also critical. We prefer dry pine needles, which we keep in a sack. However, many other materials can also be used as fuel, including rolled cardboard, hessian sacking cut into small pieces, cotton, wood shavings, dried leaves, bark or straw.

Smokers go out when not given a few puffs every few minutes. While inspecting your colony, remember to glance at your smoker to check if it is still smoking. If not, a few puffs will usually relight the fuel. If the smoker goes out, relight it with a barbecue lighter or butane torch. Also, check that there is fuel left. Smoking fuel can quickly be used and must be replenished, particularly if you inspect several hives. When the smoker is burning well, pack as much fuel as possible into the smoker to avoid that problem.

1. Use a butane torch to light a smoker and relight it quickly if the fuel burns out.

2. Place dried leaves, pine needles, cardboard or other material into the smoker.

3. Use the butane torch to quickly light the smoker, and add more leaves when the fuel starts to burn.

CHAPTER 3

Looking After My Equipment

Getting ready

Should my hives be painted?

Problem: **My hives are made of wood, and I am concerned they will rot quickly if left without protection.**

Solution

External wooden hive parts need protection from the weather; otherwise, they will quickly rot. Any external household paint can be used, although we prefer latex or water-based paint since it is less toxic to bees if the inside is painted. Also, latex paint dries quicker than oil-based paint, and painted boxes can often be replaced on a hive the same day if painted during warmer, dry weather. An alternative to painting, frequently used by commercial beekeepers, is to 'hot-wax-dip' the hive part. The hive part needs to be unpainted during hot-wax-dipping, since the hot wax needs to be absorbed by the pores in the wood to provide protection. Hot-wax-dipped hive parts last longer than painted parts, although hobby beekeepers may experience difficulty finding a person with the necessary equipment and expertise. Hot-wax-dipped hives are typically pre-assembled and painted while still hot so that the paint absorbs in the wood. This procedure is dangerous, requires specialist equipment and is generally outside most hobby beekeepers' scope.

In North America, boxes are frequently made of Western red cedar,

Row of hives in an apiary.

Above: Artistically decorated hives.

Right: A stencil used to identify hives uniquely.

Douglas fir or sugar pine. Due to their natural rot resistance, Western red cedar and Douglas fir do not need painting, although staining will bring out the natural grain in the wood and make the hives look more attractive. However, if your hives are to last, both sugar pine and soft pine used in other countries need to be painted or hot-wax-dipped.

Boxes made of polystyrene must also be painted using a water-based latex paint. Oil-based paints cannot be used on polystyrene since they dissolve the plastic. Many beekeepers use masonry paint to paint polystyrene and wooden hives because it is more durable and chip-resistant than house paint. Boxes made of rigid plastics need not be painted.

Traditionally, hives are painted white, although many beekeepers also paint with various colours. Bees do not appear to have any preference for the colour of the hive. Many hives consist of multiple boxes, each with a different colour. If you keep two or more hives near each other, painting each a different colour or pattern will assist returning bees in identifying their colony so that one colony does not get full of bees while another empties out. Some beekeepers with artistic talent decorate their hives with a variety of patterns, and these also look good in an apiary.

Many countries require the registration number to be prominently displayed on the hive by branding or painting it on each box.

Should the brood box be the same size as my supers?

Problem: I rarely move the brood box while performing inspections, while I frequently move supers. Supers full of honey are heavy. I want to use different box depths for super and brood boxes.

Solution

Mixing hive boxes containing different numbers of frames (i.e., eight- and ten-frame boxes) in the same hive is, while possible in an emergency, not advisable (see 'How do I manage my hives with boxes of different sizes?'). Therefore, when we talk about sizes in this context, we mean 'depth' rather than frame numbers. It is, however, possible, and some would argue it is advisable to use boxes of different sizes (depths) within the same hive. Many beekeepers use brood boxes and supers of various sizes. There are advantages and disadvantages to this. All boxes should be the same size for ease of management when interchanging hive boxes. Supers full of honey, however, can be heavy, and many people prefer to use smaller-sized boxes that are easier to lift. Our experience is that we like all boxes to be the same size since it is inconvenient to replace a box to find that spare boxes are different and the frames we are using will not fit in them. To facilitate removing boxes, it is possible to transfer some of the frames out of the box into an empty box so that the supers are lighter.

Although this is difficult to generalise since naming and box heights vary considerably between countries, some beekeepers use 'jumbo frames,' larger than full-depth frames that fit the depth of two stacked shallows. This way, the two shallows are combined in one large 'brood box' containing eight or ten large frames.

An eight-frame super is placed over a ten-frame brood box. A strip of wood covering the gap compensates for the difference in size.

How do I manage my hives with boxes of different sizes containing different numbers of frames?

Problem: I have two hives with boxes containing different numbers of frames (eight frame and ten frame), and I want to merge them into one.

Solution

Using hive boxes containing different numbers of frames, say an eight frame box and a ten frame box, can be done temporarily in an emergency using a piece of wood cut to cover the size difference, but this is far from ideal and should be avoided. One way to achieve this is to use a box with more space and add empty frames with comb or foundation to make up the difference.

What do I do when hive boxes and the inner cover are stuck together?

Problem: **Especially when bees are crowded, they will occasionally either build comb between the top of the frames and the inner cover or use propolis (i.e. sticky resinous material collected from plants by bees) to glue the inner cover to the top of the frames. I have experienced this problem and wonder how best to deal with it.**

Solution

In this situation, prevention is better than cure. Ensure the bees are not crowded and the top box is full of honey. Boxes full of honey generally have fewer bees than boxes containing brood. To prevent bees from attaching the inner cover to the top of the frames, try using a flexible material such as linoleum or sackcloth directly on top since these flexible materials are easier to detach from the top of the frames than a rigid inner cover. Make sure there is a 2.5 cm gap (1 inch) around the material so the bees can access the space between the flexible material and just under the lid. Bees need to patrol this area so that pests such as small hive beetles can't get established.

To detach the inner cover from the frames, use a hive tool at a corner to prise open a small gap. Gradually move the hive tool around the perimeter of the hive box until the inner cover is detached. It helps to open the hive periodically and scrape away any comb and propolis that might be present between the inner cover and the top of the lid. With a flexible cover, detaching it from the top of the frames is quickly done by peeling from a corner.

What design of bottom board should I use?

Problem: There are several types of bottom board on the market, and I need to decide which one to use.

Solution

Bottom boards can be divided into two categories: ventilated and non-ventilated. Ventilated bottom boards allow debris to fall from the hive directly to the ground, while non-ventilated bottom boards have a solid base, so the bees must remove any debris that falls at the bottom. It is, therefore, important that the hive's opening is level with the bottom board so that the bees can easily sweep out debris (generally only an issue with nucs that have an entrance above the bottom board).

Ventilated bottom boards have either a mesh or slots at the bottom allowing debris to fall to the ground. This results in a cleaner hive and

Above: A solid bottom board.
Left: Mesh bottom board.

reduces the impact of small hive beetles and *Varroa* mites. A ventilated bottom board also allows more air to flow through the hive, resulting in a generally easier-to-cool hive in the summer. In the winter, the ventilation allows for a drier hive (i.e., less condensation) but makes it harder for the bees to keep warm. Some ventilated bottom boards are fitted with a slot underneath that allows inserting a solid board, effectively closing the ventilation. In this case, it is essential to clean this board regularly as debris will accumulate there. Because it is inaccessible to the bees, it will quickly become a breeding ground for wax moths and small hive beetles if they are in your area.

What should I do when bees enter the hive through cracks and holes?

Problem: **Over time, brood boxes and supers will decay, and small gaps may appear between hive boxes, generally where hive tools have been used, and paint has been damaged. Bees will quickly start using these small gaps as shortcuts to different areas in the hive. I wonder whether this is a problem and how I can deal with it.**

Solution

Having unintended multiple entrances to a hive is not a good situation, and, in some jurisdictions, such as Australia, this is illegal. The main reason for that is that hives with multiple (unofficial) entrances are much more prone to robbing (see 'What should I do if my hive is being robbed by neighboring hives?'), which can spread diseases between hives. A short-term solution is to fill the gap with used newspaper or tape to prevent the bees from using their 'unofficial entrance'. However, the defective box must be replaced or repaired as soon as practical. Indeed, the bees, having been used to this unofficial entrance, tend to want to use it again and will try to reopen that gap. Replacing the defective box is quickly done by removing it, placing a new box on top of the hive where the old box was, and moving each frame into the new box, maintaining the orientation and sequence of frames. Shake all the remaining bees into the new box. Depending on the damage present and the woodworking skills of the beekeeper, the damaged box can be repaired or just replaced.

What should I do when the comb turns hard and dark?

Problem: The comb bees use to make brood will turn dark, hard and somewhat brittle after several seasons. I am wondering what I should do with this comb.

Solution

Once bees use comb for many life cycles, the wax becomes hard, dark, and somewhat brittle and the comb becomes less suitable for brood production. The change in comb colour results from many residues from bees hatching, while the cleaning performed by the young worker bees does not remove all these residues. A dark comb can also harbour some pathogens and pesticides the colony has encountered over the seasons. For these reasons, it is essential to rotate the frames and provide a fresh foundation or a starting strip every few years (see 'Should I use foundation and if so, should it be plastic or beeswax?').

To rotate the dark frames out of the hive, the frames mustn't contain any brood. This can be achieved by placing the frame above the excluder for approximately three weeks when any brood present will have emerged. If the comb contains drone brood, make sure to allow the drones to escape as they can't pass through the excluder and will eventually die, leaving a layer

Black comb that is old, brittle and needs replacing. It is better not to extract honey from black comb as it may be contaminated with pesticides.

of dead drones on top of the excluder that can impair worker movement into the super. If there is a strong nectar flow, the bees will have started to fill the frame up with nectar, and now, one only needs to wait until the frame is full of honey. Once capped, the honey may be harvested from the dark comb, although we prefer not to harvest honey from the black comb due to the risk of pesticide and miticide contamination. To remove the residual honey post-harvesting, place the empty frame in the hive (above the excluder) for a day, allowing the bees to clean up the mess. Do not leave the frame for longer, as the bees may now start filling up this frame with fresh nectar again. The empty, cleaned-up frame is now ready to be recycled by removing the wax (scraping if the foundation is plastic, cutting if the foundation is wax). While technically, it is possible to harvest the wax in a dark foundation, many beekeepers do not go to this trouble as the yield is low and the amount of residue is comparatively high.

Some beekeepers, especially those with many hives, expose frames to be recycled to steam that melts away the dark foundation. This process takes a lot of time to set up but is quick for treating entire boxes of frames.

CHAPTER 4

Clothing

A good bee suit makes inspection comfortable.

Many styles of beekeeper clothing are available.

What style of protective clothing should I use?

Problem: **Many protective clothing designs are available, and I must choose one that suits me.**

Solution

The cheapest option is to buy a simple veil, often preferred by commercial beekeepers. A disadvantage is that a veil provides the least protection, and bees frequently enter near the face. Although bees inside a veil rarely sting since they spend their time trying to get out, one or more bees buzzing next to your face is distracting. Although you may consider buying a veil as a backup to be used when family or friends visit and would like to see inside your hive, we do not recommend a veil for a beginner; instead, an overall or jacket would be our advice.

Protective jackets are inexpensive and come in various designs; skimping on this essential piece of equipment is unwise, as protection from bee stings is a priority. Some are designed as a vest and pulled over the head, while others include a zip so that it is like a jacket and is easier to put on. An advantage of the jacket design is that you can leave the zip open when not inspecting the colony, which is convenient, particularly on hot days. If the jacket is fitted with a zip make sure that there is a flap to protect the small gap that might exist at the end of the zip when the zip is closed. Indeed, any small opening will be discovered by bees, particularly if the hive is aggressive. Since a jacket with an attached veil only protects the upper body and head, the beekeeper must wear loose jeans with thick socks covering the bottom of the trouser legs. Bees have a nasty habit of stinging unprotected ankles and an even worse habit of climbing, undetected, to the crotch area before making themselves known. By this time, it is difficult to remove the bees, particularly if you are inspecting an open hive.

Attached veils come in different designs, and it is important to select a jacket design that allows the veil to be removed for washing. Another feature of a pullover protective jacket is that the front of the hood can be

Many styles of protective clothing are available.

A veil can be used with gentle bees — overalls with more aggressive bees.

Different styles of protective clothing are available. The choice depends on the personal preference of the beekeeper.

While more expensive, an overall provides the best protection. A jacket worn with thick jeans and socks also offers good protection. The best jackets and overalls can be zipped open at the front.

A veil is inexpensive but provides less protection than a jacket or an overall.

unzipped and pulled over the head, providing better visibility and less physical restriction. Having a hood that allows opening is also convenient for regularly taking a sip of water as dehydration is a known hazard to beekeeping.

An overall provides the best protection, although at a higher price. When buying a jacket or an overall, ensure that there are elastic bands or loops on the wrist and ankle cuffs. The elastic on the wrist is pulled over the thumb, ensuring the sleeve does not move up the arm, exposing the wrist to bee stings. The elastic at the ankles is pulled under the foot to stop the trouser legs moving up the leg, exposing the ankles. Another essential feature is elastic or Velcro around the cuffs, resulting in a tight seal between the cuff and the wrists and ankles. These can be added later if your protective clothing does not include elastic straps to pull over the thumb or under the foot.

What beekeeping clothing is best for hot climates?

Problem: The weather where I keep my bees is hot or humid, and wearing protective clothing is uncomfortable.

Solution

Protective clothing can be made from many kinds of fabric. Many suppliers manufacture protective clothing of several layers of aerated, breathable fabric. These are much cooler to wear since they allow air to flow through the fabric. Some protective clothing material is thin, mainly clothing bought online, and, although cool to wear, offers little protection from bees stinging through the fabric. It should be avoided.

When selecting a jacket or overall, look for a design that includes zips allowing the veil, or head cover, to be taken off and a zip on the front like a jacket. This will allow the beekeeper to remove or loosen clothing during hot weather or when the bees are less likely to sting. Each end of the zip needs to be fitted with a Velcro flap to cover the small gap that often exists at the end of a closed zip.

In hot climates, aerated protective clothing provides good protection and keeps the beekeeper cool. Jackets and overalls made of this material can be purchased. The material is generally made of three layers: a fine mesh inner and outer layer of nylon, and a layer in the middle made of a wider mesh of flexible material similar to rubber.

An aerated jacket is cooler in hot weather.

What should I do if bees get inside my veil?

Problem: **When I wear a veil, bees get inside, which is disconcerting.**

Solution

Bees inside a veil are more disconcerting than a threat. Bees trapped inside a veil rarely sting and are more concerned about getting out. Usually, the bee or bees concentrate around the mesh at the front and can often be squashed by hand. It is essential that the beekeeper remains calm and moves away from any flying bees. Then, slowly remove the veil, allowing the bees to escape. If you see a professional beekeeper with a bee inside their veil, notice that the beekeeper is not disturbed and usually waits till the bee lands on the mesh before squashing her. If you get excited and move your head rapidly, this will alarm the bee, increasing the likelihood of being stung.

Does wearing standard beekeeping gloves make hive inspection difficult?

Problem: **Beekeepers wear bulky gloves to protect their hands and wrists against stings. Although this is effective, bulky gloves minimize the beekeeper's ability to perform finer handling, such as manipulating the queen.**

Solution

Although thick beekeepers' gloves are imperative when handling aggressive colonies, colonies are often gentle and may be inspected with less protection of hands. As an alternative, we frequently wear nitrile gloves since they are thin, provide high security, and allow the beekeeper to handle a queen safely. The only time we have been stung when wearing Nitrile gloves is when a bee is trapped between two fingers, and we accidentally close the gap between them, pushing the stinger through the Nitrile. When a colony is disturbed, the bees become agitated and likely to sting. Experienced beekeepers handle frames slowly and gently, without disturbing the bees, and rarely get stung. As you gain experience as a beekeeper, practice handling hive parts slowly and gently and observe how the bees remain calm. This is the best way to protect against stings to the hand, allowing more flexible forms of hand protection. Judicious use of smoke to calm the bees and move them away from frames to be taken out also helps.

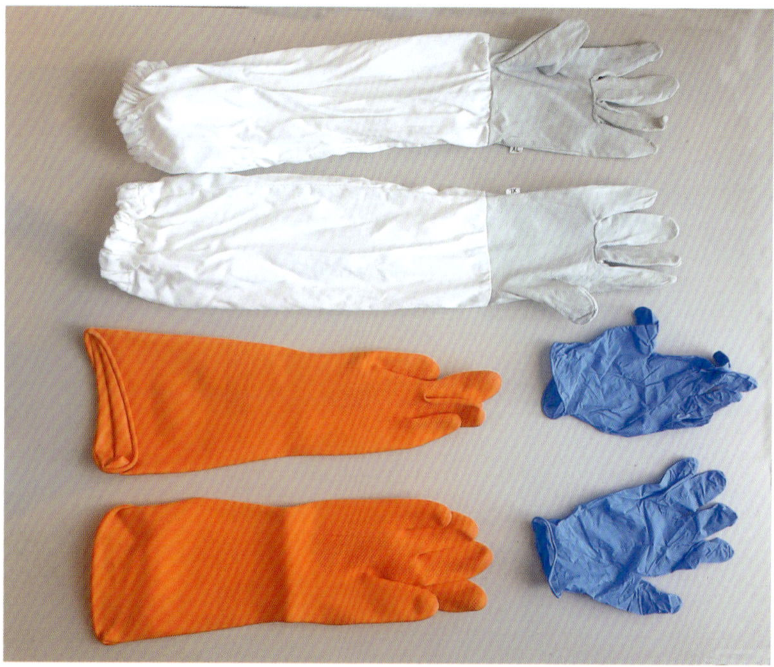

Beekeeping gloves: Cowhide, calf hide (top) or similar gloves provide good protection against stings. Some gloves include aerated wrists to keep hands cool. Nitrile gloves (bottom right) provide almost as much protection as leather gloves but are more flexible and are helpful when handling the queen. Heavy chemical or garden gloves (bottom left) also provide good protection.

CHAPTER 5

Locating Hives

A row of hives

Locating your first hive may need some thought.

Where can I place my hive?

Problem: **I need to find a suitable location for my new hive, and I'm wondering what to consider when deciding on a location.**

Solution

New beekeepers want to place their hive in a location suitable for the bees, beekeepers and neighbours. Unfortunately, these requirements are often conflicting, although, with a bit of thought, a satisfactory compromise is usually available.

The direction of the hive's entrance is important, and in the northern hemisphere, it should point towards the southeast (northeast in the southern hemisphere) so that the entrance faces the morning sun. Bees need to start collecting nectar and pollen early in the day, and ensuring the entrance to the hive is warm will encourage forager bees to leave and collect a good store of food for the colony. Also, keeping the entrance warm, particularly during colder months, will assist in keeping the hive free of condensation and fungi.

If you plan to place your hive on bricks or a stand, ensure these are in place before locating the hive so the hive does not need to be moved. Also, ensure that all long grass around the entrance is kept short, allowing foragers easy access to the hive. This is particularly important during spring when grass and weeds proliferate. Another consideration is to ensure the floor of the hive slopes gently forward about three degrees. This will stop water collecting on the base and help keep the inside dry.

A critical consideration is your neighbours. The entrance to the hive should not be close to your neighbours' fence, with the entrance pointing at their barbecue or children's play area. Some countries have specific rules about the distance between hives, fences and fence heights; check with a local bee club or bee equipment supplier whether that is the case in your area. The hive should also be placed away from public footpaths to prevent interfering with people walking past. Many experienced beekeepers ensure their hives are kept out of sight of neighbours and pedestrians. Some misguided people believe bees are a public nuisance and when they become aware of a local beekeeper, they may make the beekeeper's life difficult by complaining to them or the authorities, even if there is no risk or inconvenience to anyone.

Depending on the area, adequate shade may be needed to prevent the colony from overheating during the summer heat, while during winter, the hive should have access to the sun's warm rays all day. These requirements may be challenging, and some beekeepers move their hives between summer and winter depending on the climate.

See 'How can I move my hive to a different location?'

Should I keep one or several colonies?

Problem: As a beginner, I must consider how many hives I will keep. I want to manage several hives, but I am unsure if this is a good idea in the beginning.

Solution

Most experienced beekeepers recommend, space permitting, starting with two colonies. Keeping two hives for the first few years will teach you the skills and resources needed to manage a colony. Also, if you have two colonies, you can inspect them both and understand that both colonies, although healthy, can look very different on the inside. If you live in a large town or city, the size of your property will dictate how many hives you can keep. A helpful rule of thumb is three hives for each ¼ acre. If you have a ½ acre lot, you can keep up to six hives, and so on. Also, keeping one nuc (see 'What equipment do I need to perform additional hive manipulations?') for every two or three hives is good practice. Indeed, if a colony dies, absconds or needs additional bees to keep it strong, this nuc is ready. During swarming season, the number of hives on your property may increase as you manage any swarms. However, be careful not to keep swarm colonies for long periods as you may soon have too many colonies on your property and become unpopular with your neighbours and the local authorities. If you want to expand but do not have room to keep more hives, look for nearby empty spaces or ask other people if you can keep bees on their property. Many people have heard that bees are in trouble health-wise and would like a hive on their property without the responsibility of looking after them.

Can I keep bees in a city?

Problem I live in an urban area with little free space for bees to forage, but I would still like to keep bees.

Solution

Keeping bees in urban areas is a popular hobby for many people. The only requirement is to have space for a hive in a location where the bees will not disturb your family or your neighbours. Although this sounds counterintuitive, colonies in urban areas generally are healthier and are better fed than many colonies in rural areas. This is because metropolitan areas usually have year-round parks, river areas with plants on their banks and homes with well-tended gardens where foragers access a reliable food

source. Colonies in rural areas often have limited food supplies dictated by the widespread use of monocultures' insufficient food sources during winter or other periods of the year, depending on location, climate, and geography.

Even a small backyard can house a hive, and many urban beekeepers keep their hives on the roof or in an elevated position. One of the authors once successfully kept six hives on the 6th-floor roof of a hotel in a major city. The colonies thrived due to the ready access to year-round flowering plants in the nearby botanic gardens, along the river, and the many parks in the downtown area. Most homes in the suburbs include well-tended gardens; even homes with neglected gardens include many weeds supporting nectar-producing flowers. Large towns and cities also provide a ready water supply in hot, arid regions, even during droughts.

From top: Not too aggressive bees can be kept in urban areas. Remember to provide a source of water.

Flow Hive next to a window where they can conveniently be observed.

Hives can even be kept on rooftops.

How can I keep bees in a public space?

Problem: **I want to display my bees to the public and I would like a hive to be in a public space while at the same time minimising the risk to the public.**

Solution

Initially developed in France by Abeille Avenir, the entrance to the chimney hive is repositioned about three metres above ground level, by attaching a chimney at the hive's entrance. All other aspects of the hive and its management remains the same. Different types of hives can be used including a Flow or Langstroth hive. Since the bees leave the hive well above the heads of any people in the area, the bees do not disturb the public even if they come close to the hive. Windows, covered when not in use to preserve the darkness inside the hive and maintain temperature insolation, allow the public to observe the bees inside the hive and on the landing board. Provisions need to be made to allow cleaning of the landing board as bees cannot carry dead bees up the chimney. The bees mostly walk up the side of the chimney rather than fly within the chimney. Due to the height of the chimney, which provides substantial leverage, it is essential to fix this type of hive securely to the ground using, for example, a concrete slab to prevent it from tipping over when the chimney is being pushed.

The chimney hive is also a great solution if you have a small backyard and want to use it for other activities, such as entertaining friends. It makes for a great conversation starter and will delight children who can observe the bees directly.

In most areas, the chimney hive design conforms to regulations about separating bees from neighbours, but check with local authorities prior to proceeding. Indeed, in many areas, a solid fence is required between the hive entrance and neighbouring properties. In this case, it could be argued that the entrance is the landing board and that the bees escape well above the height of the fence.

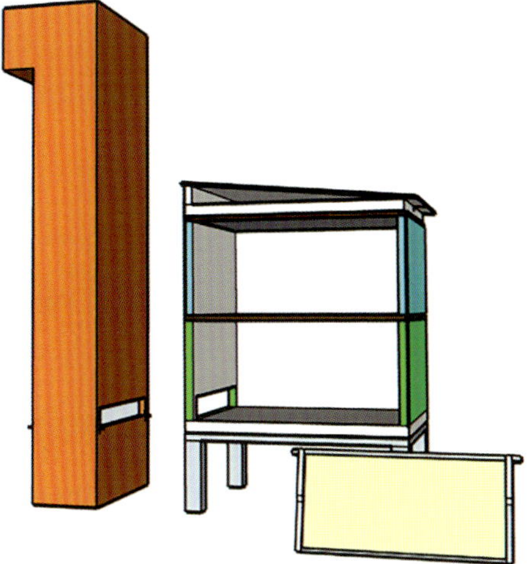

A chimney hive allows bees to be kept in a public space.

Diagram of a chimney hive viewed from the front and the back.

A chimney hive installed at the University of Melbourne, Australia.

CHAPTER 6

First Hive Opening

Plenty of bees and honey.

What can I learn about my colony from observing the hive entrance?

Problem: I want to accurately infer what is happening inside the hive by observing what is happening at the hive entrance. This would allow me to know whether I need to open the hive and address problems or whether I should leave the hive alone to prosper.

Solution

A beehive is very predictable and goes through cycles dictated by seasons and flowering patterns. By combining what happens at the hive's entrance with the time of year and what we observe from the environment around our hives, we can interpret this information to determine what is happening inside the hive without having to open it. This skill is best learned by observing the hive entrance for a few minutes before each hive opening and

Beginner beekeepers practise inspecting a hive for the first time.

Much can be learned about the colony by observing a hive entrance.

combining this information with the seasons and flowering patterns. Over time, correlations can be made that allow experienced beekeepers to anticipate problems or be reassured that all is well without opening the hive.

Observations at the hive entrance include the number of bees flying in and out at any given time and weather conditions. For example, it would not be unusual for no or very few bees foraging when the weather is cold and wet (i.e., winter). Still, this observation raises the alarm on a sunny spring day when plenty of flowers are around and the temperature is sufficient for wearing a T-shirt. This is unless the hive has recently been moved a short distance, for example, during a split and the foragers have returned to the original hive.

Dead bees at the entrance should be considered because they may indicate exposure to pesticides or disease. In autumn, many drones are expected to be expelled and die outside the hive entrance. Bee heads and abdomens without the thoraxes might indicate that the hive was attacked by wasps (see 'How do I deal with wasps or hornets attacking my hive?').

The presence of a laying queen can be inferred from foragers bringing in a lot of pollen. Watch for workers entering the hive with full pollen baskets on their hind legs. Indeed, without a laying queen, the hive has very little use for pollen, as the primary purpose of pollen is the supply of protein required to produce brood. Thus, in all seasons except winter and, in some areas, during a dry summer, the absence of foragers bringing in pollen should be considered suspicious and prompt an inspection of the hive to ensure that the queen and fresh eggs are present. This is particularly true if neighbouring hives in the same apiary bring in plenty of pollen.

Other signs that can be observed at the hive entrance include bees performing a maiden flight, which is a sure sign that plenty of workers are ready to start foraging. Maiden flights often occur in the middle of the day and are characterized by several dozen bees hovering mid-air, facing the entrance and circling the hive. During a maiden flight, bees will not fight at the entrance to the hive. Fighting is a sign that robbing is occurring. Bees taking maiden flights could look like robbing, which also involves bees circling the entrance, but they are easy to distinguish because the robbers frequently fight with guard bees at the entrance, while 'maiden-flighters' will not (see 'What should I do if my hive is being robbed by neighbouring hives?'), and the bees do not leave the hive in large numbers (a sure sign of swarming; see 'My colony is swarming; what should I do?'). The beekeeper does not need to do anything but enjoy the spectacle and relish that all is well with this hive. Make a mental note that the hive is likely to expand soon, mainly if this occurs in late spring or early summer. The beekeeper may also want to delay opening the hive until this process is completed so as not to disturb the foragers' memorisation of the hive entrance and the immediate surroundings of the hive.

How can you safely take out and manipulate a frame of bees?

Problem: I need to remove and inspect frames with minimal disturbance to the bees in the most efficient way.

Solution

Experienced beekeepers have learned to remove and handle frames that support many bees without disturbing or exciting them. They do this by remaining calm and ensuring each movement is slow and gentle. Using this method, any bees on or next to the frame do not become agitated and aggressive and are less likely to sting. In addition, when manipulating comb from top bar hives (see 'What are the advantages and disadvantages of a top bar hive?'), this technique reduces the risk of breaking off fresh comb.

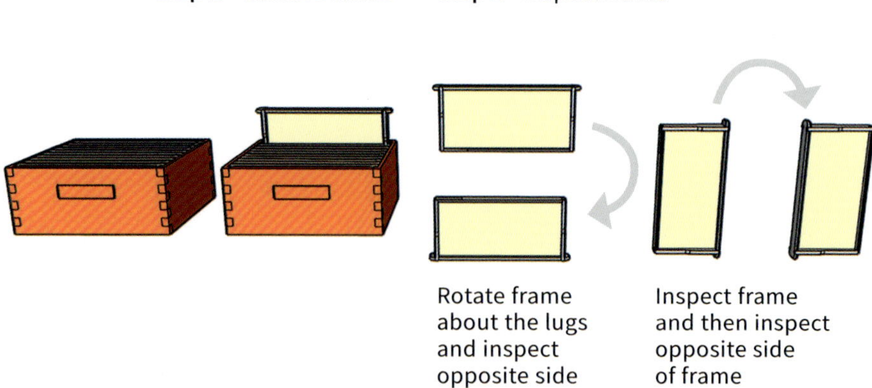

There are two methods of removing frames for inspection. Slowly and gently remove the frame from the box. Hold the top bar lugs firmly and inspect one side of the foundation. Next, rotate the frame 180° and examine the opposite side. Return the frame to the box or place it on the lid or in a spare box while inspecting the remaining frames.

To inspect a frame, hold it by the end lugs on the top bar. Slowly rotate the frame so you can see the opposite side. Usually, we rotate the frame vertically and then slowly rotate it so we can inspect the opposite side. Avoid knocking the frames against other hive components.

When removing frames for inspection, start with the frame next to the end frame. End frames tend to be heavily propolized and can be difficult to remove. If you are extracting and need to remove all frames containing honey, you can either replace each frame with a new one or extract that day and return the old frames to the super. If you are inspecting the brood, remove the first frame, inspect it, and place it on the lid next to the brood box. Using your hive tool, push the second frame away from the third frame so it is not stuck, and remove it for inspection. After inspecting the second frame and confirming that the queen is not on it, either place it next to the first frame on the lid or return it to the brood box slightly away from the next brood frame. Next, prize the third frame away from its neighbour, remove and inspect it. Each frame must be removed gently and slowly to avoid disturbing the bees.

To inspect a frame, hold it firmly by the end, top bar and lugs. Rotate the frame so the top bar is vertical. Inspect the side nearest you, then rotate the frame by 180° and inspect the opposite side. Now, return the frame to the box.

How do I safely handle bee boxes?

Problem I want to safely handle heavy supers without risking back injuries.

Solution

On average, a ten-frame deep super of honey weighs about 34 kg (75 lb), while an eight-frame deep super weighs about 27 kg (60 lb). Many people find supers full of honey challenging to lift. There are three main ways to alleviate this:

Cleats the entire width of a box, making lifting easier.

Use an eight-frame hive instead of a ten frame hive. Two fewer frames reduce the weight by 7 kg (15 lb), which may be sufficient to make it easy to move. This is one of the main reasons some beekeepers use eight frame boxes.

Use half-depth or shallow-sized supers. This will significantly reduce the weight of a full super, making them easier to lift. A disadvantage of this method is that honey is more challenging to extract from smaller frames, plus there are more frames to extract since two half-sized supers are needed to replace a full-sized super. Also, a manual extractor may need to be modified to accommodate smaller frames (see 'What are the different types and sizes of hive boxes?' for comparing box sizes).

The way a super is lifted is important. Incorrect body position while lifting even a relatively light super can lead to a back injury. First, though, attach cleats to each super for ease of lifting, do not rely only on grips cut into the super as these are often not convenient to use. Secondly, posture is also essential; lift the super with your legs while maintaining a straight back. While lifting the box, lift the box at an angle, with the edge facing your body. Lifting a super with the flat side facing you can be much more difficult.

CHAPTER 7

Spring and Summer Management

Checking the hive.

What do I look for during the first inspection in spring?

Problem: I am about to perform my first inspection in spring and wonder how I should go about it.

Solution

The last winter inspection of your hive, or even just watching the bees come and go from your hive, should prepare you for what to expect during your first spring inspection (see 'How do I prepare my hives for winter?'). This first spring inspection is a crucial responsibility for every beekeeper, setting the tone for the upcoming season. Examine your hives in spring when the colony has stopped clustering and bees are out flying. Do not open your hives if the air temperature is below 18°C (34°F), which may kill many of the young brood and weaken the colony.

Open the hive on a warm, sunny day and thoroughly inspect every box and frame. This is particularly important after the long winter months when such inspections were impossible.

During this inspection, you need to consider the following:

- **How strong is the colony?**
- **Are eggs or brood present? If they are, you know the queen is there even if you do not see her.**
- **Is the brood pattern patchy or solid?**
- **How much food is left if the weather deteriorates and the bees cannot forage?**
- **Are diseases such as American foulbrood, European foulbrood and Chalkbrood present in the hive?**

Procedure for the first inspection of spring

Remove the lid and place it upside down on the ground next to the hive. Place the brood box on top of the upturned lid; if two brood boxes were

Hives at a commercial apiary.

used over winter, place both boxes on the upturned lid. If you use a pitched roof as a lid, place the box on two parallel pieces of wood spaced to align with the box's edges.

Now that you have complete access to the bottom board or hive base, do the following:

- **Clean debris off the bottom board and wash it with a wet cloth.**
- **If an entrance reducer is used, it doesn't need to be removed during the first inspection but can be removed a few weeks later.**

If you used two brood boxes over winter, it's time to inspect and clean each one. Consider reversing them by placing the previous top brood box onto the base. Over winter, the colony may have moved to the upper brood box and will not utilize the total volume of the two brood boxes. By reversing the position of the two brood boxes, you will break up the winter cluster, giving the colony plenty of room for the queen to lay her eggs.

When inspecting the brood box, be methodical and follow the steps outlined below.

- **With the first brood box on the base, remove each frame and check it for brood, honey and pollen stores.**
- **If brood is present, note if there is a solid capped or uncapped brood pattern. If the brood pattern is patchy, this may mean disease, an infertile queen or a lack of food stores, and immediate action is required.**

- Check to see that the queen is present. If you cannot locate her, look for newly laid eggs or very young larvae to confirm her presence.
- If you find a queen, mark her as described in 'Should I mark my queen?'
- If you decide the colony needs re-queening, it is usually too early to buy a new queen in the season, and this activity will probably need to be performed later in the spring or early summer, when queens can be more readily obtained.
- If you replace frames in a hive during early spring, preferably use frames with drawn comb kept over from the previous year so that the queen can start laying eggs in them immediately without the delay of the workers first needing to draw out the comb.
- If there are empty frames between the brood and honey frames, place the frames containing honey next to the brood. However, be careful not to divide the brood by placing the frames of honey between the brood frames, effectively splitting the brood into two parts.
- If the weather has been poor, check the hives weekly for food. If insufficient food is present, start feeding immediately.

ADDING A NEW SUPER

Clean the hive mat and lid.

- Clean the hive mat.
- Remove and discard any burr comb from the inside of the lid. See photograph opposite.
- If the inside of the lid looks damp, additional ventilation is needed. When you place the lid on the brood box or super, raise it slightly by placing matchsticks, small twigs, stones or iceblock sticks under the corners. This will allow damp air to escape from the hive, keeping the inside drier.

Clean the area around the hives.

- **Control weeds and vegetation around the hive and next to the hive entrance. Wear protective clothing if you use an electrical or motor-driven device to cut the grass.**
- **Bees require a lot of water at this time of year, so make sure it is available.**
- **Remove rubbish around the hive and clean and tidy the apiary area.**

An essential aspect of hive inspection is swarm management. Many books and beekeepers say that to perform swarm prevention correctly, the brood box must be inspected every spring for queen cells. A weekly inspection disturbs the colony too much and seriously interferes with their regular nectar and pollen collection activities. The bees can become so agitated that you may be unable to go near the hive for several days without getting stung. Later in this chapter, we will describe other methods of swarm management, and if these are followed, a detailed inspection of your hives need only be performed approximately every three weeks.

Where possible, avoid vibrations in the direct area around the hive, such as using a lawn mower or edge trimmer. This is true even if you are not inspecting the hive since the noise and disturbance may result in bees attacking nearby people.

Burr comb is the comb that workers build under the lid. Burr comb can contain either honey or brood.

When do I add a second super or brood box?

Problem: **Depending on the time of year and the colony, you may need to add either a second brood box or an additional super. Later in the year, you may need to reduce the size of the hive.**

Solution

Adding a second brood box: During late spring and early summer, when the queen is laying many eggs (up to 1500 per day), and the amount of uncapped and capped brood is increasing, you many need to add a brood box. Indeed, if about 70 per cent of the cells in the brood box contain brood, you need to add a second brood box, allowing the colony to rear more larvae and expand further.

When adding a second brood box, remove the super and queen excluder and place the new brood box containing frames above the old brood box. A good practice is to include some brood frames with larvae in the new brood box; this encourages the nurse bees and queen to move to the new brood box earlier than they may otherwise. In doing so, keep the brood together and as much as possible in the middle of the boxes, with the empty frames on the outside. Indeed, the aim here is to simulate a natural brood nest where the brood forms a sphere in the middle of the hive. Place the excluder on top of the new box and replace the supers above the excluder.

Removing a second brood box: Later in the year, when less food is available and the queen lays fewer eggs, resulting in less brood, the second brood box needs to be removed. Removing a second brood box is more difficult than adding a brood box. Removing a brood box requires planning and time; a second queen excluder is also needed. First, remove the supers and queen excluder. Remove the top brood box and place a second queen excluder between the first and second brood boxes. Now replace the top brood box with the excluder above it, beneath the supers. Wait a week or two and check both brood boxes. The box that contains the queen will also contain eggs and young larvae; the box that does not include the queen will only contain capped brood cells. Place the box that contains the queen, eggs and larvae on the bottom, next to the base, place an excluder above it, and place the brood box without the queen and supers above the excluder. Over the next few weeks, the capped brood in the top brood box will emerge, and the box can be removed. Alternatively, a patient beekeeper may locate the queen visually and ensure that the queen is in the bottom box before moving the queen excluder between the two brood boxes. After a few weeks, the eggs in the top brood box will have emerged, and a few weeks later, the bees will have used the space to collect honey, which can be harvested and the box removed from the hive (especially in preparation for winter, see 'How do I prepare my hives for winter?'). Always keep drone brood below the excluder as emerging drones are too large to fit through the excluder and a layer of dead drones will form on top of the excluder hindering movement of worker bees.

Adding or removing a second (or third) super: Adding or removing a super is easier than for a brood box since the queen and larvae will not be found in supers.

To add a super, remove the lid, inner cover or hive mat and place a box of frames above the top super. Some beekeepers prefer to put the new super above the excluder on the brood box, below the old super, since worker bees are sometimes reluctant to move into a new super, mainly if it is on top and has undrawn frames. Placing the new super as the bottom super overcomes this as workers need to move through the new super to work on the old super, quickly getting used to the new super.

Removing a second (or third) super is easier than removing a brood box. Remove the top super and place it next to the hive. Remove each frame and shake/brush the bees into the remaining super. When the super has been cleaned of bees, it can be taken away and stored for winter. Some beekeepers with many hives use a leaf blower or fume board to remove the bees from the entire box (see 'How do I remove bees from my supers when harvesting honey?').

If the frames still contain honey, you have several options to remove or store the honey. One method is to add a second queen excluder above the top super and replace the super to be removed above this. Next, use a scratcher or hive tool to break the wax surface of every cell containing honey, and replace the frames into the top super to be removed. Over the next few days or a week, worker bees will usually clean the honey from the damaged cells in the top super and repack it in the bottom super. After a week, inspect the top super; if all the honey has been cleaned out, remove the super and store it over winter (see 'How should I store my excess frames?').

If the frames contain a lot of honey, you can either store the full frames for later extraction or use them to provide supplementary feed to your bees during winter. If you decide to store frames of honey over winter, they must be kept in a sealed container that will not allow beetles, moths or other creatures to enter (see 'How should I store my excess frames?'). Before storing, the frames need to be frozen so small hive beetle and wax moth eggs are destroyed.

How do I add a new super?

Problem: My hive is full of bees, how do I add another super?

Solution

If the colony is sufficiently strong and congested with bees, add a super on top of the brood boxes and, if used, a queen excluder between them. A sign that a super needs to be added is that the brood box is full of honey and brood, or when you remove the lid, bees overflow the sides of the brood box. The super being added should also preferably only contain drawn comb since we are coming into the main honey flow period, and the bees should not spend time drawing out a new comb but should instead store nectar.

Often bees will not move up into a new super to store nectar. To overcome this, place some brood frames in the new super, and the nurse bees will move up to look after the brood, encouraging other workers to follow.

How do I manage the build-up of bees in my hive to maximise honey production?

Problem: About one to two months before the start of the main flowering period, I want to increase the amount of brood to maximise honey production during the honey-flow period.

Solution

The number of eggs a queen lays is a function of the amount of nectar and pollen collected. The result is a delay between the start of the main nectar-collecting season and the peak number of foragers in the colony. To overcome this, feed syrup and pollen/pollen substitute to the colony starting about one to two months before you expect nectar-producing plants in your area to bloom. The queen, sensing the availability of food, will begin laying eggs before the availability of nectar-producing flowering plants. Pollen/pollen substitute must also be provided since this provides proteins essential for the development of larvae. See also Chapter 9, 'Feeding bees syrup or pollen/pollen supplement'.

How do I perform a routine summer inspection?

Problem: I want to perform routine hive inspections during late spring and summer and wonder how best to do this.

Solution

The season's second and subsequent hive inspections are much more reduced than the first inspection, as explained in 'What do I look for during the first inspection in spring?' About every three weeks or less, open the hive and check for the following:

Beekeeper inspecting his hive. Gentle bees mean that he need only wear a veil, but no gloves.

Inspect the supers for honey

- Remove the lid and hive mat and inspect some of the frames in the super to determine how much honey is present. If most frames are two-thirds (or more) full of capped honey, they can be removed as they are ready for extraction, although many beekeepers leave the filled frames in the hive and harvest once or twice a year. If the amount of capped honey is much less than this, determine how strong the colony is and how strong the honey flow is, to give you an idea of when to inspect next for capped honey ready to extract.
- Clean any burr comb from the box or frames.

Inspect and clean each brood box

- Every second or third inspection, check in the brood boxes for a laying queen. If you cannot locate her, look for newly laid eggs or very young larvae to confirm her presence.
- If brood is present, check that the capped and uncapped brood patterns are solid. If the brood pattern is patchy, this may mean disease, an infertile queen or a lack of food stores, and immediate action is required.
- The above two checks should tell you whether the colony needs re-queening.
- If you replace frames in a brood box, use frames with drawn comb, where possible.
- If the weather has been wet and/or cold, check the hives weekly for food. If insufficient food is present, start feeding immediately (see Chapter 9, 'Feeding bees syrup or pollen/ pollen supplement').

Adding a new super

- If the colony is sufficiently robust and has a good honey flow, you can add another super to the existing super. The new super should contain some drawn comb so the colony

can start storing nectar immediately.
- Workers often do not like moving into a new super. To help the bees get used to the new super, try placing some frames of honey or even some capped brood in it. Placing the new super just above the excluder below existing supers also helps (see 'When do I add a second super or brood box?').

Clean the lid

- Remove and discard any burr comb from the inside of the lid cavity.
- If the inside of the lid looks damp at the start of spring, additional hive ventilation is needed. When you replace the lid on the brood box or super, raise it slightly by placing matchsticks, small twigs, stones or iceblock sticks under the corners. This will allow damp air to escape from the hive, keeping the inside drier while preventing bees from entering or leaving the hive (additional entrances promote robbing).
- Condensation inside a hive arises from bee metabolism, which keeps the brood warm. If the weather is cold, ventilation is counterproductive if it makes it harder for the bees to thermo-regulate the brood. If the weather is extremely hot, add additional ventilation by raising the lid slightly using matchsticks, small twigs, stones or iceblock sticks under the corners. The lid should only be raised somewhat for the duration of the heat wave.

Clean the area around the hives.

- Control weeds and vegetation around the hive, particularly next to the entrance.
- Bees require a lot of water during the summer; ensure that it is available.

FEDSS — FOOD, EGGS, DISEASE, SPACE, SWARMING

A helpful mnemonic to remember when inspecting hives is FEDSS: Food, Eggs, Disease, Space, Swarming.

- **Food: are there sufficient stores of honey and pollen present?**
- **Eggs: if you cannot locate the queen, but there are eggs or very young larvae in cells, the queen was present a day or two before.**
- **Disease: check for the signs of disease.**
- **Space: is there sufficient space in the hive for the queen to lay eggs or for the workers to store honey?**
- **Swarming: if the colony is congested or queen cells are present during spring and early summer, you should split the hive into two colonies.**

There are few bees in my hive; should I worry?

Problem: I noticed fewer bees in my hive during an inspection than I expected. Should I leave the colony alone and hope they will increase in number, or should I take some other action?

Solution

A colony with fewer bees than expected is not unusual. This is a normal part of the colony's natural cycle and may not be a cause for concern. Over winter, when less food is available, the queen stops or reduces the number of eggs laid, resulting in fewer bees entering early spring. Also, if food becomes scarce during late spring or summer, the queen may again reduce the number of eggs laid. During early spring, as food becomes available, a lower number of bees in your colony means that less honey will be produced. Many beekeepers overcome this by providing supplementary food in early spring or late winter. Syrup with a 1:1 sugar:water ratio by weight should be used, and supplementary pollen or substitutes should also be provided. Use a sugar:water ratio of 1:1, not 2:1, since 1:1 encourages the queen to lay more eggs, while 2:1 encourages the colony to produce honey.

Comparing both colonies is a good practice if you have more than one colony. Each colony has characteristics of its own, including the number of bees. Still, it is good to learn firsthand that each colony is unique and to appreciate the differences that may occur. These differences are in part determined by genetics. While some bee strains start to build up numbers early in spring, focusing on increasing numbers of workers, others seem to accumulate more nectar before increasing the worker population. Depending on the local environmental conditions, different strains may perform quite differently. Similarly, the propensity of bees to swarm differs greatly depending on genetic factors.

While continuous supplementary feeding may be necessary in areas with insufficient natural food, it's important to note that beekeeping may not be practical in such conditions. A colony cannot thrive solely on supplementary feeding, as natural nectar and pollen provide essential vitamins and trace chemicals for a healthy colony.

If the queen is old or not laying enough eggs, it's crucial to replace her. Requeening every year or two with a queen from a reputable source is a good practice. For more information, see 'What strains of queen should I buy?'

A diseased colony or queen may also be the cause. Check the health of your colony during frequent inspections and take appropriate action as needed. Some pathogens like chalkbrood may be reduced by requeening with a disease-resistant queen; other diseases may be complex to manage, and, in the case of American foulbrood, depending on the regulations in force in your country, the colony may need to be destroyed, and the equipment irradiated or burned (see Chapter 19, 'Other Pests and Diseases').

In all cases, apart from obvious signs of disease, I would first try feeding with syrup and wait a few weeks to see if the colony size increases.

How do I remove stuck frames in my super?

Problem: **The first frame in the super is difficult to remove, and I need some tricks to make it easier.**

Solution

The first frame is often difficult to remove because of two possible reasons:

- **The super is full of 'slightly too wide' frames for the box. This means the last frame must be forced into the last remaining space. It is usually the end frame that is removed first, and this is often jammed partly due to its size but also because the wood may have expanded inside a damp hive.**
- **Bees may have glued the frame with propolis, or it contains too much-capped honey.**

Either way, you can force the end frame to be removed, although it may break the top bar's lugs. Often, there is an inner frame, generally the second frame from the end, which is less likely to have the queen on it than more central frames, which typically contain more brood, that is easier to remove. When you have removed one frame, the remaining frames are easy to remove by using your hive tool to lever them away from each other, breaking the seal that is holding them together.

An alternative method is to remove an end frame and replace it with a spacer board, often called a dummy board. A spacer board is a top bar slightly thinner than the other frames, with a sheet of wood inside the frame where the foundation would generally be. Because the frame is slightly thinner, it is less likely to become stuck. Although using a spacer board means less room for the colony to store honey, inspection and removal are much more manageable.

See 'How can you safely take out and manipulate a frame of bees?'

CHAPTER 8

Autumn and Winter Management

Sunset over the apiary.

How do I prepare my hives for winter?

Problem: It is autumn, and I want to prepare my hive for winter.

Solution

It is advisable to prepare early in autumn for the winter months, aiming for the following:

- Strong hive with plenty of bees and a young queen laying well.
- A lot of stores of honey and pollen. How much you need will depend on the local area and the expected winter severity. Check with the local bee club or equipment supplier. Aim for at least one deep box of honey and some pollen.
- A hive that is packed without a lot of free space at the start of winter. This is because the number of bees and the space taken by brood will reduce as winter progresses. Empty frames will only consume energy as they need to be kept warm.

Beehives in snow.

Depending on your area, the autumn months might bring in some honey, but you can't count on that to have sufficient stores. Thus, monitor your hive regularly and provide additional nutrition early. Indeed, providing sugar water (2:1 sugar to water by weight) will build up stores, and the sugar will be mixed with nectar, making for a much better store than sugar alone. The ratio of 2 parts of sugar for 1 part of water will promote storage of the sugar solution. In contrast, a 1:1 ratio by weight promotes brood formation, which is not necessarily advisable in preparation for winter unless there are insufficient adult bees to keep the hive warm (see 'Should I provide syrup for my bees?').

The number of bees required for successful overwintering depends on the severity of the winter conditions, but most of all, on the size of the hive at the start of winter. A small hive can be overwintered in a single nuc box in many areas, but the same number of bees in a double deep will likely fail to survive. The act of reducing the available space during autumn is called packing down. It is an important activity generally performed after the last harvest of honey for the year or just before the advent of cold weather.

What should I do if my hive is being robbed by neighbouring hives?

Problem: I want to prevent my hive from being robbed by bees from other colonies.

Solution

Most robbing of colonies occurs when nectar supplies are scarce and other honey bee colonies are desperately looking for food. This typically happens in autumn, when the wasps are also often out in strength. The wasps can rob the hive not only of honey, but also brood and adult bees (see 'How do I deal with wasps or hornets attacking my hive?'). In some areas, one of the most significant honey flows may occur during late autumn; thus, robbing is less of a problem. To restrict robbing, you should:

- **Reduce the hive's entrance to 5 to 10 centimetres (2 to 4 inches) wide. Bees are better able to defend the smaller entrance from intruders.**
- **Make sure there are no secondary entrances through cracks and degraded parts of the hive boxes.**
- **Ensure that the hive is left open for the minimum amount of time when you inspect it. Consider covering the exposed frame tops when opening the hive.**
- **Do not leave burr comb, frames of honey or spilled syrup near the hive to attract robbers.**
- **When near your hives, watch the entrance for unusual activity. Robber bees are very fast and aggressive in their attempts to enter the hive. If you notice this type of activity, shut down the entrance to the hive entirely for an hour, and the robber bees may go away.**
- **If you suspect robbing may occur in the following day or two, it is essential to act immediately to minimize the risk.**
- **If robbing is an issue in your area, place a robbing screen over the entrance of weak hives. A robbing screen allows the house bees to find their way in and out of the hive through experience, while the robbing bees will go directly to the area where the smell of honey is greatest and hit the mesh protecting the entrance.**

Apart from mass robbing, which only occasionally occurs, robbing attempts by individual intruders are an everyday occurrence. Watching a hive entrance often shows bees pushing other bees away. This pushing away seldom involves stinging, but the intruder soon gets the message and moves on. Potential robbers being pushed away from the hive often have different body markings and colours than the colony they are trying to penetrate.

See also 'What should I do if wasps or hornets attack my colony?'

A robbing screen confuses robbers and allows workers to enter and leave while limiting access by robber bees. The entrance to the hive is through the small hole at top right of screen. Robbers detect the scent of honey and concentrate at the screen, not realising the entrance is elsewhere.

Robbing guards are available on-line or at most beekeeping supply stores. They may also act as mouse guards.

Can my hive survive in the snow?

Problem: I want to keep bees in a climate with snow or extreme cold.

Solution

Some beekeepers, particularly in parts of North America and northern Europe or at higher elevations, live in climates that include snow or extreme cold during winter. With the correct preparation of the colony and hive in late autumn, your colony can survive and emerge in the early spring, ready to start rearing brood, collecting food and producing honey.

If the temperatures in your region fall below -7°C (20°F), you need to leave 40 to 55 kg of honey (90 to 120 lb) for the colony to survive. If there is less than this amount of honey in the hive towards the end of autumn, you must provide syrup with a 2:1 mixture of sugar and water by weight during warm weather for the colony to build a sufficient store of honey.

Since cold weather is often intensified by wind, provide a windbreak on the north side (in the northern hemisphere) to provide shelter. Also, locate your hive so it is in the sun during the day to get the maximum warmth. Windbreaks can be made of shrubs or a fence about 1.5 metres (five feet) high. The boards in the fence need not be solid but must be sufficiently close to stop cold winds from reaching the hive.

An alternative, often essential if the temperature is frigid, is insulating the hive with sheets of polystyrene, corrugated plastic or commercial hive covers such as Bee Cozy and Eazy On. Many options are available; check with your beekeeping supply store, club or other local beekeepers.

The following guidelines can be used when preparing your hives for extreme winters:

- **Remove the two frames on the farthest two sides of the brood box and replace them with dead boards. Dead boards are frames with 1 cm (3/8 inch) thick wood in place of the foundation. This will provide additional insulation on two sides of the hive.**

Insulating a hive is an excellent way to ensure the colony survives a harsh winter. Snow does not usually mean cold for long periods. Hives covered with snow for short periods can be kept warm by bees.

- Restrict the entrance to the hive to minimize the entry of cold air.
- If external insulation is used, ensure that the entrance to the hive is not covered and that air can enter. If there are two entrances, one at the bottom and the other at the top, both entrances may need to be clear.
- Use a slotted base-board to provide additional insulation at the bottom of the colony.

- Place a hive box containing insulating material such as straw, shredded paper, polystyrene, fibreglass mats or wood shavings under the lid and on the brood box. A box under the lid to provide insulation is often called a quilt. This will also help keep the hive dry by reducing condensation.
- Ensure the hive is slightly above ground, and leave a dead air space under the hive.
- Mice love to make nests in hives over winter. Place a mouse guard over the entrance to stop them from gaining entry.
- Ensure that there is adequate ventilation to stop the build-up of humidity.
- If large amounts of snow are expected, consider having the entrance in the top part of the hive close to the lid, to prevent the entrance from being obstructed by snow.
- In areas where winter temperatures fall below -20°C (-4°F), many beekeepers move their hives into a shed or other insulated room. This is particularly important if cold weather lasts longer than five months.

How should I store my excess frames?

Problem: Excess frames removed during preparation for winter, often called pack-down, must be stored for reuse in spring.

Solution

In preparation for winter, the size of the hive is usually reduced, resulting in at least one box and eight or ten frames that need to be stored for several months. If the frames contain honey, the excess honey can either be extracted or the frames can be stored as discussed in Chapter 14, 'Harvesting honey'. The frames can be left in a shed, garage or outside if they do not contain honey. The difficulty with storing empty frames that

A frame is stored in a plastic bag for winter.

contain beeswax foundation is that they attract wax moths, which can destroy a set of frames within a few weeks, resulting in a mass of tangled matting produced by the larvae as they devour the wax comb.

Wax moths do not like light, so one way to store frames is to leave the hive's lid off so light can enter. Alternatively, you can store frames that do not contain honey in plastic bags or sealed plastic boxes that prevent moths from entering. Before storage, these plastic bags with the frames must be frozen for several days to kill any eggs in the comb.

CHAPTER 9

Feeding Bees Syrup or Pollen/ Pollen Supplement

Pouring sugar syrup in a feeder.

What is good nutrition, and how do I ensure my bees are well-fed?

Problem: I want to ensure my colony has adequate nutrition based on local flowering plants. If this is insufficient, I can supplement by feeding syrup and pollen/pollen substitute

Solution

The health of your colony depends, to a large extent, on providing adequate nutrition. Although nutrition can be complex, it can be broken down into simple categories of essential nutrients:

1. Carbohydrates: essential as an energy source, carbohydrates can be supplied in two forms.

- Nectar is the primary source of carbohydrates and is obtained when worker bees collect nectar from flowers and convert it into honey for long-term storage.
- Sugar syrup is an alternative supply of carbohydrates if insufficient nectar is coming in. Although syrup is suitable as emergency food and is often used to increase colony strength, it does not contain the range of vitamins, minerals and other honey chemicals essential for bee health. When feeding syrup, be sure to only use pure white sugar rather than raw sugar, which is not as well tolerated by bees.

2. Proteins: essential in the development of larvae and the maintenance of adult bee health.

- Pollen is the primary source of proteins and is obtained from pollen that has been partly digested or fermented by bacteria and enzymes to produce bee bread for long-term storage in the hive.
- Pollen substitutes are an important alternative source of proteins and are often supplemented with a range of vitamins and minerals. Pollen substitutes are also deficient in many

Feeding pollen substitute to a colony.

substances essential for bee health and should only be used when pollen is in short supply.

3. Lipids are fats used to build cell membranes (walls) and other vital functions.

- Lipids are mainly obtained from pollen, although food supplements often also include these critical chemicals.

4. Vitamins and minerals are obtained from pollen and to a lesser degree from nectar. These are essential for overall bee health and the proper functioning of many biological processes.

- Pollen is the primary source of vitamins and minerals, and nectar is a secondary source. Food supplements often include these nutrients.

5. Water is needed for various physiological functions, including digestion, thermoregulation and hydration.

- Water can be provided from various sources, including natural, beekeeper-supplied and domestic sources.
- When supplying water ensure that the bees have a place to land near the water source (floating on the surface or rocks protruding from the surface). Otherwise, the bees will drown while collecting water.
- In many areas it is a legal requirement to provide water for

the bees so that they do not source water from neighbours, causing a nuisance (see 'Should I provide water for my bees?').

TIPS TO ENSURE BEES ARE WELL-FED

1. Ensure a diversity of flowering plants.

- Include a wide variety of flowering plants that bloom during different times of the year, including winter, for a continuous supply of nectar and pollen.
- Ensure that there are native plants and garden plants to provide a diverse source of nutrition. In areas where honey bees are native, they often prefer native plants over garden plants as they sometimes provide better nutrition.

2. Provide supplementary food during times when food is scarce.

- Sugar syrup can be fed most of the year, from early spring to late autumn or fall. Use a ratio of 1:1 sugar:water by weight during spring to stimulate comb formation and egg laying and 2:1 sugar:water during autumn to stimulate the accumulation of stores.
- Pollen or pollen-substitute patties must be provided when natural pollen is insufficient.

3. Water

- Ensure a supply of clean water, although foragers often seek out salty swimming pool water, or even standing water, which is believed to be due to the presence of minerals.

4. Hive weight

- Regularly check the weight of each hive. This is particularly important during winter when you cannot open the hive for inspection. An unexpected fall in weight may indicate a health problem with your colony.
- The hive's weight also reflects the presence of honey and pollen. If necessary, provide supplementary feeding until sufficient stores of food are available.

Focusing on these will help you ensure that your bees are strong, are healthy and have sufficient food for their health and productivity.

How do I know if the bees have enough food without opening the hive?

Problem: I want to check that my colony has sufficient honey to last the winter or other unfavourable periods for foraging, but conditions do not allow me to open the hive.

Solution

If the weather is favourable, check how much honey is available. You need not even remove frames; look down the gap between honey frames to see if many capped honey cells are present. This method is sufficiently accurate to provide a reasonable estimate of available food.

If the weather is poor or cold during winter, and you do not want to open your hive, an easy method to estimate the amount of honey is to lift one end of the top super about one centimetre (1/2 inch) above the box below. If you practise this a few times when you open the hive during better weather, you soon learn to estimate food stores by the weight of the super.

Some beekeepers place an eye screw in the side of the bottom board opposite the entrance to which they attach a spring scale and record the weight of the hive over time. The method will not give a total weight as the hive hinges around the front cleat, but the weight is consistent and can be compared over time. Some manufacturers produce electronic scales that allow remote monitoring of the hive weight over the mobile phone network or via Wi-Fi connection, warning the beekeeper about changes in his/her hive's weight when they are in remote locations. This can save considerable travel as beekeepers can reduce the number of visits to their remote apiary.

If you believe there is a food shortage, you may need to provide supplementary syrup and pollen. If performed quickly, this can usually be provided under the lid and hive mat or inner cover (see 'Should I provide water for my bees?', 'Should I provide pollen or pollen substitute for my bees?', 'Should I provide food supplements for my bees?', 'Should I provide syrup for my bees?' and 'Should I provide probiotics for my bees?').

Failing this, the bees will starve (see 'I see dead worker bees stuck head-first in the cells. What is the reason my hive has died?'). This typically occurs in spring, when many resources are needed to raise brood, reserves are depleted after winter and a cold spell prevents the bees from collecting food.

The summers in my area are hot and dry; will my bees survive?

Problem I want to keep bees in a region with long, arid summers, often with few flowering plants and little water.

Solution

Many beekeepers successfully keep bees in hot, dry areas. As long as there are flowering plants for part of the year to give the colony time to store honey for winter and to collect pollen during the spring buildup when there is little food available, it can usually be managed successfully. Always ensure a water supply is nearby, monitor the amount of honey and pollen in the hive and provide supplementary feeding as necessary. Keep the hive well-ventilated and in the shade. Hopefully, sufficient honey will be produced during the months when flowering plants are available to provide you with enough honey to make the effort worthwhile. See also 'Should I provide water for my bees?', 'Should I provide pollen or pollen substitute for my bees?', 'Should I provide food supplements for my bees?', 'Should I provide syrup for my bees?' and 'Should I provide probiotics for my bees?'

PROVIDING COLONIES WITH WATER

1. A bowl or pail of water with stones or twigs for bees to land on.

2. A chicken feeder used to provide water.

3. An inverted bottle in front of the hive.

Should I provide water for my bees?

Problem **There is not much water where I live, and I have been told I should provide water for my bees.**

Solution

Water is usually available in most locations, particularly in urban areas. The difficulty is that the water may not be nearby, and the colony needs to fly some distance to access it. During hot, dry weather, water needs to be provided next, or very close, to the hives. Lack of water in rural areas may not lead to foragers disturbing neighbours. In urban areas, however, thirsty bees may find that next door's swimming pool is the most convenient water source, significantly upsetting neighbouring swimmers. Once bees form the habit of drinking at next door's pool, it isn't easy to wean them off it.

The best option is placing a water source near the hives before a water shortage so the colony gets used to drinking from it.

There are many ways to provide drinking water to the colony. Some sources, such as water bottles, can be attached to the hive, while others are nearby. Water sources that can be placed nearby include chicken feeders and containers of water with twigs or polystyrene floating on top where bees can land. Providing floating material is essential to prevent bees from drowning. Even a dripping tap can be used as a permanent water source. Water for your bees on your property is legally required in many areas.

Should I provide syrup for my bees?

Problem **My colony has few honey reserves, and its stores do not build up over time. I need to feed sugar to build up my colony's strength.**

Solution

Most beekeepers feed syrup to their colonies sometime during the year. Usually, this occurs during the spring buildup when flowers are scarce, and the colony is desperate for honey (and pollen) to feed larvae. An adequate food supply is critical as the colony needs to increase colony numbers rapidly in preparation for a nectar flow and provide energy to foragers to seek food sources. Another time when feeding is required is when nectar is in short supply; this can occur during drought conditions or when food sources are scarce. Alternatively, suppose the beekeeper has taken too much honey from the hive near winter. In that case, the colony often uses supplementary feeding to increase honey stores in preparation for winter. This however should be avoided if possible. Finally, supplementary feeding would be essential to managing these situations when purchasing a package of bees, housing a swarm or keeping a nuc with a new queen. Indeed, during these times the colony needs to rapidly build comb, which takes a lot of energy and therefore requires a lot of sugar water or nectar. The best way to determine if a colony needs supplemental feeding is to check inside the hive for adequate honey storage. Even though a nectar flow is ongoing, with plenty of nectar being collected, beekeepers need to be alert to unexpected drought and a shortage of food. Supplementary feeding may be needed unexpectedly. Always use highly refined sugar (white sugar) and do not feed raw (or brown) sugar to bees, which is detrimental to their health.

Feeding syrup: There are many ways to feed syrup and many designs of feeders available online or for sale at beekeeping supply stores. Some designs can be installed outside the hive, making topping up with syrup easier.

Mix a cup of white cane sugar with water to make syrup, forming a 1:1 mix by weight. Sugar is highly soluble in water, and the amount that can be dissolved often astonishes new beekeepers. Sugar at that concentration will quickly ferment. It is therefore important to ensure that the amount of sugar water fed to the bees can be consumed within a few days by the colony (i.e., –1–2 liters (~0.25–0.5 gallon) at a time is generally okay, depending on the strength of the colony). Indeed, fermented sugar water will sicken the bees and weaken the colony.

Since feeder designs and uses vary considerably, details on each design will be left to the beekeeper to research. Many web-based resources provide detailed information on the advantages and disadvantages of each type.

A simple and inexpensive feeder is a Ziplock plastic bag filled with syrup. The bag is placed on the top bars of the hive, and a sharp knife is used to slit the top of the bag. Very little, if any, syrup falls out, and the bees can easily feed on the syrup through the slits. Ensure that you do not cut through the bottom of the bag, or most of the syrup will escape rapidly.

Open feeding, where sugar water (or especially honey) is left outside for any bee to access rather than being directed to a particular hive, is prohibited in many parts of the world due to the risk of spreading bee diseases. Irrespective of the legal requirement, using open feeding is a bad idea as the beekeeper risks infecting the entire apiary should a disease break out during this feeding period.

Feeding fondant: While syrup is sugar-water in liquid form, an alternative, often used in North America, is to feed fondant, a highly concentrated

1. Plastic bags containing syrup can be placed under the lid, resting on the hive mat
2. Bags are slit, giving bees access to syrup.
3. Frame feeders are a popular way to feed syrup.
4. Twigs or mesh are placed in the frame feeder giving workers a ladder into the syrup.
5. Candy fed to a colony.

form of syrup, in which the concentration of sugar is sufficient to make the water/sugar mix possess the consistency of a fondant. Use four parts by weight of white cane sugar with one part water. Heat the mixture on a stove, and constantly stir until all the sugar has dissolved. When the sugar has dissolved, let the mixture cool to 95 °C (200 °F). When the liquid temperature is 95 °C, pour the concentrate into a food mixer or stir vigorously by hand to dissolve any remaining sugar crystals. After a few minutes, when the mixture has turned light-coloured and smooth, pour the concentrate into moulds ready to be placed in hives. Fondant can be stored in grease-proof paper and kept in a cool place or in the refrigerator or freezer for extended periods. Fondant, in contrast to sugar water, does not ferment and can, therefore, be used as a long-term energy source for the bees.

Placing a large lump of fondant under the lid of the hive can minimize the risk that the bees will run out of stores during winter or early spring. This is often done in areas with sustained long and cold winters.

Another popular method to feed sugar is to dissolve sufficient sugar in water that the mixture solidifies when it cools; this is often called candy. To make candy, boil about 7.5 kg of white cane sugar (16 lb) in three cups of water. When the sugar has dissolved and the solution is clear, pour the mixture into a large mould (an aluminium baking tray is suitable for this), and let it cool. When cool, it should have solidified into a block. If the mixture does not harden when cooled, reheat it, add more sugar, and let it cool again. You will be amazed at how much sugar can be dissolved in three cups of water. Sugar candy can be fed to the colony as a solid by placing it under the hive mat or top cover. Large pieces of candy may be broken down for storage.

Should I provide pollen or pollen substitute for my bees?

Problem: During colony inspections, I noticed that little pollen was stored in the hive, and I wondered whether I should feed supplementary pollen or use a pollen substitute.

Solution

Many beekeepers feed syrup but neglect to feed pollen, which is equally important when food is limited. Historically, beekeepers used to mix pollen or yeast with syrup, make a patty, and leave the patty under the top cover or hive mat. With the appearance of the Small Hive Beetle, *Aethina tumida* (see 'What should I do when my hive is infested with small hive beetles?'), this method has fallen out of favour since the beetles feed on the patties and thrive. Many beekeepers still make patties, and a good recipe for them is to mix dry pollen or pollen substitute powder with an equal amount of sugar. To bind the mixture into a patty, add 1:1 sugar:water

The best pollen substitute for bees is rich in protein and made with ingredients like brewer's yeast, soybean flour and dried skimmed milk.

by weight of syrup to make a paste-like consistency. Patties can be kept in the freezer until needed.

Many naturally occurring substances, such as yeast or soy flour, can be used as pollen substitutes. Alternatively, commercially made Ultra Bee, MegaBee and AP23 are popular with beekeepers.

Apart from patties, pollen or pollen substitute can be added as a dry powder. Small hive beetles are less likely to be attracted to a dry powder, so it is safer to use if colonies in your area are infested. The powder can be left on grease-proof paper above the brood box or under the top cover. Alternatively, we sometimes brush pollen into an empty brood comb in the brood box, conveniently located for nurse bees to access.

Should I provide food supplements in addition to syrup and pollen substitutes for my bees?

Problem: Should I feed my bees food supplements in addition to syrup and a pollen substitute?

Solution

In addition to pollen or pollen substitutes, many beekeepers feed their colonies food supplements such as Honey B Healthy, Complete or Nozevit. If you believe your bees lack essential foods unavailable in syrup or pollen/pollen substitute, you may want to provide them with supplements that contain vitamins, polyphenols, minerals and amino acids that may be in short supply to the colony.

Should I provide probiotics for my bees?

Problem: **Many companies advertise probiotic food supplements to improve bee health. I wonder whether I should feed probiotics to my bees.**

Solution

Research in this area is ongoing, and the claimed benefits of probiotics have not been proven in field trials, so there are still questions about the benefits of probiotics to colony health. Initial laboratory studies suggest, however, that probiotics may be a promising tool for improving honey bee health. To date, the focus of supplementary feeding has been syrup, pollen or pollen substitutes, and supplementary food additives. In addition, many scientists and beekeepers believe that using probiotics, similar to yogurt, in our diet may also benefit bees. These include:

Enhanced gut health: Probiotics may improve the balance of microorganisms in the honey bee gut, promoting digestion and nutrient absorption.

Increased resistance to pathogens: Certain probiotic strains may help bees resist common pathogens and diseases by competitive exclusion (outcompeting harmful microbes for resources) or by producing antimicrobial substances.

Improved immune function: Probiotics may stimulate the honey bees' immune system, making them more resilient to environmental stressors and diseases.

Better overall health and vitality: Healthy gut microbiota may contribute to well-being, strengthening colonies, improving brood development and increased honey production.

Mitigation of pesticide effects: Some probiotics have been shown to mitigate the adverse effects of pesticides on bees, potentially offering a natural solution to protect bee health in agricultural settings.

CHAPTER 10

My Bees

Bees at the entrance of the hive.

How can I recognize different castes of bees?

Problem: **I need to know how to differentiate between workers, drones and the queen within a colony.**

Solution

There are three castes of bees in a typical honey bee colony: (i) workers, (ii) drones and (iii) the queen. The workers are the most abundant cast, typically representing over 90 per cent of the bees present. Workers are female bees and, as the name implies, are doing the vast majority of the work required for the hive to thrive. Workers can be identified by the fact that they are the smallest bees present, typically narrower than a drone and shorter than a queen. The tasks performed by workers include cleaning cells, feeding larvae with royal jelly, storing nectar and maturing it to honey, building comb,

Beekeepers often become very attached to their bees.

DIFFERENT CASTES OF BEES: Worker, queen and drone bees. Note that the length of the queen's wings does not reach the end of her abdomen, unlike the wings of a drone and worker that are almost as long as their abdomen.

cleaning the hive from debris and dead bees, collecting nectar and pollen, feeding the queen and protecting the hive from aggressors.

Drones are male bees and have only one main task, which is to fertilize a queen from another colony, after which they die as part of the fertilization or mating process. Drones are haploid, meaning they have only one set of chromosomes because they are born from an unfertilized egg. Thus, drones only contain maternal genetic material. In contrast to workers, drones are usually free to enter any hive and use its resources. They cannot sting and, therefore, do not protect the hive. They are generally only tolerated in the hive during spring, summer and early autumn. At the end of autumn, they will be barred from entering the hive and, unable to feed themselves, will die a short time later. Drones can be recognized by their larger size compared to workers, rounded abdomen and huge eyes (needed to spot the queen in flight for mating).

The queen is recognizable by her larger size, pointed abdomen and slower walking movement on the comb. One way to recognize the queen is to observe the wings, which should not be as long as her abdomen. Beginners often mistake drones for the queen, but a glance at the massive eyes of the drones and the pointed shape of the queen's abdomen with the wings well short of the abdomen's tip should make for a rapid differentiation. However, spotting the queen amongst the thousands of workers is challenging for most beginners and, occasionally, more experienced beekeepers (see 'How can I find the queen?').

What should I do if my bees are aggressive?

Problem: Keeping aggressive bees is illegal in many parts of the world and detracts from the enjoyment of keeping bees. However, it is not unusual for a hive to become aggressive over time, mainly if a new queen that has mated with feral drones has been produced. The beekeeper needs to address this quickly.

Solution

Beekeepers need to be vigilant about their colony's aggressiveness and deal with it without delay upon identifying an aggressive colony. The issue will only worsen as the hive grows and should not be ignored.

A sure sign that a hive has become aggressive is that workers will rapidly fly into the veil of the bee suit, sting any exposed skin, and seek the beekeepers after they move away from the hive for, say, more than 10 metres (30 feet). At each hive opening, the beekeeper should make a mental note about the aggressiveness of the hive. As soon as a hive becomes aggressive, the beekeeper must start replacing the queen. By replacing the aggressive queen with a queen that produces gentle workers, the beekeeper can quite rapidly solve the problem. Indeed, the typical worker will take approximately three weeks from egg to emerging, spend about the next three weeks in the hive, and the following two to three weeks as a forager. Thus, adult bees with the new queen's genetics will emerge within three weeks once a queen has been replaced. These new bees will start replacing the existing workers after that, with the first batch of new workers involved in defending the hive present about six weeks after the queen was replaced. Therefore, one can expect to see a significant effect on the aggressiveness of the hive from about six weeks post-queen replacement.

To ensure the queen possesses a gentle genetic predisposition, it is often best to order a new queen from a reputable queen-rearer. For the procedure required to replace the queen, see 'How should I introduce a new

The best way to manage an aggressive colony is to requeen with a gentle queen.

queen into a queen-less colony?' Finding a queen in an aggressive hive is difficult and sometimes unpleasant. It should preferably be done by several beekeepers, including an experienced person (see 'How can I find the queen?'). In addition, one should be aware that aggressive hives often reject a newly introduced queen and that precautions can be made to avoid this (see 'What should I do if my newly introduced queen is being killed by the colony?').

What should I do if I am allergic to bee stings?

Problem: **I am allergic to bee stings and wonder what to do about my beekeeping activities.**

Solution

There are different degrees of being allergic to bee stings, all of which can have serious consequences. Therefore, any uncertainty about allergies should be taken very seriously, and medical advice should be sought.

This is especially true since allergic reactions can become more severe over time, although the reverse is possible. Reactions to bee stings range from relatively mild local reactions typically associated with local inflammation (swelling, redness, pain and a warm skin feeling), to hives (a red, itchy skin rash at a site distant from the area directly affected by the sting), to the hazardous anaphylactic shock, which can include neurological signs (fainting, nausea, confusion (like being drunk)), respiratory signs (shortness of breath, rapid and shallow breathing), heart rate increase or decrease and even death. One of the main dangers of these allergic reactions is that they can occur and develop into a complete anaphylactic shock within minutes of being stung, leaving very little time to act, especially if the beekeeping activity happens far from medical facilities. Under these circumstances, there is a real danger, and beekeepers who know they are allergic must always have an EpiPen.

Even if you are only mildly allergic, and developing local reactions near the bee sting, multiple bee stings should be taken seriously, especially around the face, throat and neck. It is, therefore, good practice to discuss with your doctor the possibility of having antihistamines at hand. Local creams used to treat insect stings and bites while decreasing the discomfort of the sting victim (typically a reduction in itchiness and a moderate reduction in the amount of local swelling) will do very little against an allergic reaction.

Someone who has never or only once been stung by a bee can't know whether they are allergic to bee stings. Indeed, several stings, days to weeks apart, are required to mount the responses necessary to become allergic. If a person has been stung several times and develops any symptoms away from the area directly affected by the sting (say, an itching rash on the torso while being stung on the shoulder or an itchy feeling in the palm following a sting to the leg), he/she should discuss this with their doctor as these are sure signs that an allergic reaction is developing and subsequent stings could be dangerous.

Beekeepers who have developed allergies must consult their doctor before deciding whether to continue beekeeping.

If a decision is made to continue beekeeping, it might be possible to undergo a desensitisation procedure. However, this may take up to two years and require regular injections of small amounts of bee venom. Again, this should only be done under the care of a specialist in a medical setting. Any attempt to self-treat in this area may have deadly consequences and should not be undertaken under any circumstances.

What should I do if I am stung?

Problem: I have just been stung and wonder what the best approach is.

Solution

The steps to undertake are very different for allergic vs non-allergic people.

Non-allergic people: If the person has been stung several times in the past and it is known that there is no allergic reaction, the aim is to minimize the local discomfort. Unless there are many stings at once or in the face/neck/throat area, one does not have to do anything, and most beekeepers will leave things as they are. It is possible to reduce the swelling by taking some oral antihistamines. In general, antihistamines do not have severe side effects. They are generally safe (talk to your doctor), and many beekeepers will carry some antihistamine as part of their beekeeping kit. Antihistamines work well for bee stings because they perform two functions simultaneously: (i) neutralise the histamine present in bee venom and (ii) neutralise any histamine produced by the body in response to the bee sting. Antihistamines will reduce the local inflammation (swelling, redness, pain and warm feeling) at the sting site. Since they are taken orally, they disperse in the entire body (i.e., systemic) and therefore also reduce systemic effects often experienced by mildly allergic people (i.e., hives; see 'What should I do if I think I am allergic to bee stings?').

Allergic people: Dangerous allergic reactions occur generally very rapidly (within minutes) after being stung; therefore, time is of the essence. If a person is known to be allergic to bee stings and is stung by a bee, immediately contact emergency services. If an EpiPen is available, prepare it immediately. Monitor the person for symptoms, particularly changes in breathing patterns, heart rate and hives. If these symptoms occur, encourage the person to self-administer the EpiPen or, if trained and authorised to do so, administer the EpiPen. Ideally, you will talk to emergency services or a medical doctor at this stage. Be aware of your legal responsibilities and obligations, as they vary in different countries.

If there is no signs of an allergic response (i.e., no hives, no changes in breathing or heart rate) after 15 min., it is unlikely that an anaphylactic shock will still happen, although milder localized reactions such as swelling, redness, pain and warm feeling may still happen. Continue to monitor for at least one hour.

What is the difference between honey and bee bread?

Problem: When looking at frames in a beehive, I can see capped honey and cells filled with coloured substances (bee bread). I wonder what the difference is and how the bees use them.

Solution

Honey is made predominantly from sugars and is, therefore, mainly a store of energy for the colony. In contrast, bee bread is predominantly composed of pollen. The pollen is a source of proteins and lipids (fats) for the hive. The protein is essential to producing brood and developing the eggs into larvae, pupae and adult bees, while the lipids are used for building cell membranes. A healthy hive should have plenty of honey and bee bread stores.

Coloured pollen (bee bread) in a frame. Pollen comes in various colours depending on its source.

How can I tell the difference between capped brood and capped honey?

Problem: **Capped brood and capped honey look similar; how can I tell their differences?**

Solution

The differences between capped brood and capped honey are easy to identify. The wax on capped honey has a glossy, whitish appearance. The surface of capped honey is relatively flat or uneven. The wax on capped worker brood is slightly convex, protruding from the surface, like a bullet. Brood capping is also more likely to be light brown in colour and have a rougher texture. The rougher texture is a result of the capping being porous, allowing air to enter the cell so the pupae can breathe.

See photos in section: 'What is the difference between honey and bee bread?'

Capped worker brood on top has flat, brown wax capping. The capping of drone brood at bottom is more bullet shaped. Brood capping is rough and porous.

Honey has flat, glossy capping.

What should I do when there is little honey in my hive?

Problem: The hive is very light, and I observed very little honey. I am concerned that the bees in the hive might starve.

Solution

It is good practice to always keep at least the equivalent of one to two frames of honey in the hive so that if there is an unexpected weather event that prevents the bees from leaving the hive or if only a small amount of nectar is available, the hive will not starve. Starved hives are characterized by dead worker bees stuck head-first in the cells (see 'I see dead worker bees stuck head-first in the cells. What is the reason my hive has died?').

Making sure that there are plenty of supplies in the hive is particularly important and even critical in early spring when the hives are building up the number of bees and, therefore, need to have a lot of resources available. This is also the period when (depending on your location) the bees may come out of a harsh winter, which might have depleted the honey reserves. In that case, immediately start feeding the bees 1:1 sugar:water syrup by weight. This will also stimulate brood production; therefore, the amount of pollen should also be checked (see 'What should I do when there is not much pollen in my hive?').

Having large honey reserves is less critical during late spring and early summer when plenty of nectar is likely to be available. This depends on local conditions, as in some dry areas, the summer is too dry to produce much nectar. Consult local beekeepers, clubs or bee equipment retailers for local knowledge.

In early autumn, the bees must be assessed for sufficient honey for the winter months. This not only because honey is a food store, but also because frames filled with honey are an excellent thermal mass that helps keep the hive at a constant temperature. Depending on local conditions, you may need eight full-depth frames of honey to overwinter hives. If there is any chance that there might not be enough honey for overwintering, it is critical to start feeding the hive a 2:1 sugar:water mixture by weight as soon as possible while the bees still bring in nectar and pollen. Indeed, the sugar supplied will be mixed with the nectar and pollen still coming in, and this mixture will greatly favour overwintering compared to just stores of sugar water.

In an emergency, if the hive is very light in winter, it can be opened quickly for feeding. However, opening the hive should be avoided in weather colder than about 16°C (60°F), as the colony could be harmed.

If there is insufficient honey in the hive to feed the colony, frames from the previous harvest can be stored for use in emergencies.

Bees with heads in cells is a classic sign of starvation.

What should I do when there is little pollen in my hive?

Problem: I looked in my hive, and while there are good honey stores, I can see very few cells filled with bee bread. I wonder whether this is normal and what, if anything, I should do about it.

Solution

The absence of bee bread in the hive will hamper brood production, as proteins stored in the form of pollen in bee bread are an essential component for brood production. While this is not a problem in winter when only a few eggs are laid (if any), this may be critical in the leadup to spring when few flowers may be available, but the colony is rearing brood. Nevertheless, building the protein stores for when the bees need it in spring is worthwhile. Indeed, from early spring onwards, the hive will start producing a large amount of brood, requiring a lot of protein and lipids. It is, therefore, advisable to feed the hive with pollen supplement (see 'Should I provide pollen or pollen substitute for my bees?').

Even in late autumn, when the queen has often (depending on the geographical area) stopped laying, pollen reserves are needed. This is because pollen will be required in early spring for the build-up of worker bees. Hence, pollen reserves should always be present in the hive. Feeding dry pollen to a colony may be necessary if small hive beetles are a problem.

Feeding pollen substitute to a colony is explained in Chapter 9.

How do I deal with laying workers?

Problem: **In a healthy hive, only the queen lays eggs. However, workers can start laying eggs when the queen fails or dies. These eggs are often not viable, and the situation should be remedied.**

Solution

Workers are female bees and can lay eggs. However, because the workers have not mated, their eggs are unfertilized. These eggs are, therefore, haploid (i.e., they have one set of chromosomes). As a result, only drone brood can be produced this way (see 'How can I recognize different castes of bees?'). However, the workers are not experienced at laying eggs and often lay several eggs in the same cell. The presence of several eggs in the

Workers lay multiple eggs in cells two to three weeks after the queen dies. At this stage, workers are called drone layers.

same cell is one of the tell-tale signs that workers and not the queen have laid the eggs. No new workers will be produced since these eggs are haploid and generally unviable. If the queen is incapacitated or dead, the fact that these eggs are haploid, and therefore do not have the genetic material needed for a female bee, also means that the hive can't make a new queen from the worker-produced eggs. Hence, over time, the hive will invariably die out.

To remedy this situation, most commercial beekeepers shake off all the bees on frames from a worker-laying hive in front of a healthy queen hive. The now homeless bees will walk to the new queen-right hive and will, hopefully, merge with that colony.

Is it normal to observe workers of different colours at the entrance of my hive?

Problem **When looking at the entrance of one of my hives, I can see workers entering that are quite different in colour, ranging from mainly yellow to primarily black. I thought they would be more similar because workers are all descendants of one queen.**

Solution

If the same queen has been in the hive for more than three weeks, all the workers will be descendants of that one queen. However, the queen typically mates with up to 20 drones during one to four mating flights taken a few days after she emerges from the hive. Thus, most workers are sisters or half-sisters with the same mother but possibly different fathers. Hence, if the queen has mated with drones from other strains of bees, the workers will have a wide range of genetic characteristics depending on what was inherited from their father. This may include not only different colours but

1. Yellow-straw-coloured bees are often Italian bees, *Apis mellifera ligustica*. A colony may have a mix of bees with different colours if the queen has mated with drones from various strains.

2. Musky grey bees are often Carniolan bees, *Apis mellifera carnica*, originating from parts of Austria, Slovenia, Serbia and nearby areas. Carniolan bees are called carnies by their keepers.

3. Caucasian bees, *Apis mellifera caucasia*, are dark in colour and originate in the Caucasus region of Georgia and parts of southern Russia.

4. Native to much of western and central Europe, the European Dark Bee, *Apis mellifera mellifera*, is often called the German Dark Bee. Many traditional beekeepers in the UK and Europe are working to popularise this strain of bees.

also various levels of productivity, aggressiveness and disease resistance. To obtain a pure strain of bees with a more uniform genetic background, the beekeeper must proactively determine the drones with which the queen will be mating. This is not easy but can be done by, for example, flooding the drone congregation area where the queen mates with drones of similar genetic backgrounds, using isolated colonies such as on an island, where only a controlled population of drones is available, or use instrumental insemination where semen from selected drones is used. These techniques are typically used by professional queen breeders who want to control the genetics of the queens they produce carefully.

Italian bees have brown and yellow bands. Among different strains of Italian bees, there are three colours: Leather; bright yellow (golden); and very pale yellow (Cordovan). Their bodies are smaller, and their overhairs are shorter than those of the darker honey bee races.

Carniolan honey bees are about the same size as the Italian honey bee, but they are physically distinguished by their generally dusky brown-grey colour that is relieved by stripes of a subdued lighter brown colour. The Carniolan bee has a very long tongue, which is very well adapted for clover.

Caucasian honey bees originate from the high valleys of the Central Caucasus. Georgia is the 'central homeland' for the subspecies, although the bees can also be found in eastern Turkey, Armenia and Azerbaijan. Caucasian bees are noted for their long tongues, or proboscises, which enables them to extract more nectar from flowers.

Dark European Bees are native to Europe and can be found from the Pyrenees mountains in the southwest to Scandinavia and the UK. They can be broadly distinguished from other subspecies by their stocky body, abundant thoracal and sparse abdominal hair, and dark coloration. When viewed from a distance, they appear blackish or rich dark brown. They are large for honey bees, though they have unusually short tongues.

How can I predict what will happen in my hive several weeks in advance?

Problem: **I need to predict what will likely happen to my hive in the medium term, for example, when planning a holiday or period away from the hive.**

Solution

The life cycle of bees within a hive is highly predictable as they move from eggs to larvae to pupae to adults (Table 1 below). Once born, bees will start doing certain tasks sequentially at defined times depending on their caste (workers, drones or queens).

Caste	Hatch days	Capped days	Emerge days	Final activity days
Queen	3½	8 ± 1	16 ± 2	Laying 28 ± 5
Worker	3½	9 ± 1	20 ± 1	Foraging 42 ± 7
Drone	3½	10 ± 1	24 ± 1	Mating 38 ± 5

TABLE 1: For each caste of bees within the hive, the number of days it takes from laying an egg until each event.

Worker bees start by being confined inside the hive and, therefore, have no sense of where the hive is located. At the same time, they go through a cycle of sequential tasks over time (see Table 2 on p. 128). While the cycle is well-defined, it can be stretched somewhat if needed, particularly for older bees reverting to tasks that they have done previously. This would be the case when, for example, after a swarming, there is a need to rapidly produce a large amount of comb to start a new colony. Another instance is when workers doing specific tasks have been removed from the cycle. This could be, for example, when doing a split, where all the field bees have been removed due to the split being moved to a different (nearby) location. Under these conditions, some workers may start doing that task a little earlier.

Time window since birth	Tasks
House bee (never left hive)	
1 - 2 days	Cleans cells and warms the brood nest
3 - 5 days	Feeds older larvae with honey and pollen
6 - 11 days	Feeds young larvae with royal jelly
12 - 17 days	Produces wax and constructs comb, ripens honey
18 - 21 days	Guards and ventilates the hive
Field bee (knows the position of the hive)	
22 + days	Forage for nectar, pollen, propolis and water
Approximately 40 days	Dead

TABLE 2: Typical timeline of worker bee tasks undertaken after birth.

Similarly, a queen will only mate up to about ten days after birth and start laying eggs several days after mating (Table 1). After about two weeks, drones will be mature enough to leave the hive, searching for a queen to mate.

How can I move my hive to a different location?

Problem: The location of my hive is no longer suitable, and the hive needs to be moved.

Solution

While moving a hive less than three metres or more than three kilometres can be done without too many issues, moving hives between three metres and three kilometres requires great care to avoid losing many foragers.

Indeed, foragers have a great sense of where the hive is located and, under normal circumstances, will return to the hive entrance using local landmarks. If the hive has moved less than a few metres and with the entrance pointing in the same direction, they will rely on smell to find the entrance. If the hive has been moved more than three kilometres, they will no longer recognize typical landmarks and reprogramme their flight path to backtrack to the new hive location. Problems arise when foragers can recognize landmarks they have seen before, but the hive has moved. Indeed, under these conditions, they will go to where the hive entrance was and cannot find the new hive entrance location. This also happens when the hive entrance is re-oriented and points in the opposite direction of where it used to be. There are three ways to move a hive between three metres and three kilometres, effectively.

If the distance is only up to, say, 100 metres (100 yards), the hive can be 'walked' or moved one to three metres (up to one to three yards) every few days until the hive is at the new location. The hive is usually left in the same place for about three days before moving again. This slow hive movement minimizes the number of foragers that return to the original location. However, this takes a lot of time and assumes that the path is clear so that the hive does not cause any nuisance while moving along the path.

The same principle applies to rotating a hive; to change the orientation of the hive entrance, gradually rotate the hive in steps of about 30 degrees every few days. Do not rotate a hive by more than 30 degrees to allow the bees to adapt gradually to the new orientation.

Secondly, suppose it is impossible to 'walk' the hive to the new location using small steps. In that case, an alternative is to move it by vehicle to a new, intermediate location greater than three kilometres (2 miles) from the initial location. The hive is left at the new location for about three weeks. During this time, all foragers who regarded the initial location as home have died, replaced by foragers who only knew the intermediate location. After three weeks, the hive can be moved to the new, final location, assuming the new, final location is also greater than three kilometres from the intermediate location.

Many beekeepers find that moving the hive directly to the new location and obscuring the entrance with vegetation or other obstructions is also effective. Obstructing the entrance confuses the foragers and causes them to reorient themselves to the new location. While this is an easy method, it does not always work.

Lastly, it is also possible to close the hive entrance for three to four days, by which time they have 'forgotten' where the hive is and will re-orient upon leaving it for the first time. If contemplating this method, ensure the hive is in the shade and has plenty of water, food and ventilation. This method works well, especially in winter or when the weather forecast is for several days of cold and wet weather when the bees would typically not fly anyway. We have found that this lock-up method can be combined with the previous entrance obstruction method for the best results.

How far do bees fly to obtain food?

Problem: **I am concerned that there is insufficient food or water for the bees near my hive and wonder how far the bees will fly to collect food or water.**

Solution

Bees can fly up to three kilometres (~1.8 miles) from the hive to collect food or water. Thus, the bees will scout an area of approximately 28 km^2 (approximately ten $miles^2$). However, they will first collect food from areas closer to the hive before venturing further afield. Therefore, one can expect that the area explored by the bees will be smaller in times of abundant food. This is important when moving hives to different locations (see 'How can I move my hive to a different location?')

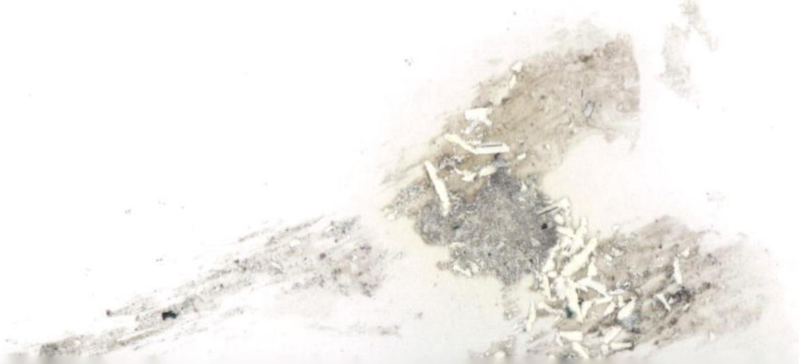

What can I do if I see many dead drones outside the hive?

Problem: **I am concerned about the many dead drones outside the hive entrance.**

Solution

In winter or during other periods when food sources are scarce, workers realise that the colony cannot support a large drone population with their limited honey stores. Since drones cannot survive without the assistance of workers to feed them, the colony forcibly removes drones from the hive, where they die outside in large numbers. This is a natural part of the colony life cycle. When food becomes available again in spring or after a drought, workers will encourage the queen to lay more drone eggs, and the drone population in the colony will increase.

In contrast, a large number of dead workers in front of the hive suggests pesticide poisoning, particularly if their tongues are hanging out. During robbing, bees don't try to kill one another, and in case of a wasp attack, the diagnostic would be an abundance of heads and/or wings without thoraces and abdomens.

I have two weak colonies; how can I merge them?

Problem: **Worker bees are generally only allowed to enter their hive and will be attacked if they enter another hive. As a result, merging colonies needs to be done with great care to prevent a deadly battle between bees from each colony.**

Place two or more sheets of newspaper between the brood boxes when merging colonies. The bees will take two or three days to remove the paper, allowing them to merge without fighting.

Solution

To avoid a fight between queens from different colonies, when merging, there must be only one queen. Indeed, if there is more than one queen in the merged colony, the queens will find each other and engage in a deadly fight. This may result in the queen that you wanted to keep being killed. Much better is to find and kill the queen in the colony that you do not wish to keep before merging the colonies. This is generally the weaker colony but could also be the colony with the most aggressive genetics.

To provide time for the bees from both colonies to get to know each other, it is best to merge two colonies by placing a sheet of newspaper on top of the first colony; pierce the newspaper with a few slits (use a boxcutter or sharp knife) and put the second colony on top. We recommend that the second colony should be the weaker colony. The bees will slowly, over the next day chew on the newspaper and remove it from the hive through the entrance. Both colonies will learn to know each other, and minimal fighting will occur.

If the size of the hive boxes from the two hives differs, say one hive is an eight frame hive while the other is a ten frame hive, closing the gaps between the two is imperative. This can be achieved by transferring the hive from the smaller boxes into a larger box and filling the space with empty frames. In our example, this would mean transferring the eight frames of the eight frame hive into a ten frame box before merging. Alternatively, one can use a core flute insert the size of the largest box (in our example, a ten frame-sized core flute sheet, with a rectangle cut out the size of an eight frame box. This effectively forms a rim protecting the frames that protrude from the larger box.

I see dead worker bees stuck head-first in the cells. What is the reason my hive has died?

Problem: **My hive has died, and I see many dead bees with their heads inside the cells. I wonder what is happening to my hive.**

Solution

A hive with many dead bees with their heads inside cells means that it ran out of food stores and has died of starvation. Indeed, in a last-ditch effort to collect the last bits of honey from the stores, the workers are licking the bottom of the cells, and when the honey runs out and insufficient nectar comes in, the bees die from undercooling. This can be prevented by always leaving enough honey for the bees and by feeding the bees early on when low levels of honey are detected (see 'What should I do when there is not much honey in my hive?'). While it is now too late to save the hive, the frames can be collected, frozen and stored for reuse later (see 'How should I store my excess frames?').

See p. 121 photograph of starving workers with heads in cells looking for food.

CHAPTER 11

Swarming

Overcrowding can lead to swarming.

How do I prevent swarming and produce more hives?

Problem: **I want to prevent swarming and expand the number of hives in my apiary (or sell hives).**

Solution

The two most common methods of preventing swarming are splitting the colony or using the Demaree method of swarm prevention. Splitting the colony is discussed in the section 'How do I split a hive', and the Demaree method is discussed in the section 'What is the Demaree method of swarm control?' Langstroth and top bar hives can be split to prevent swarming, while the Demaree method best suits Langstroth hives. If you plan to rear a second hive while managing swarming, splitting is suitable, while the Demaree method does not produce a second hive.

A swarm landing on a tree.

What is the Demaree method of swarm control?

Problem: The Demaree method minimizes swarming while maintaining the same number of hives after the manipulation and maximises honey production. Compared to splitting the hive, the Demaree method also conserves heat, as the brood is kept at the top of the hive, which is less prone to chilling the brood if an unexpected cold spell occurs.

Solution

The essential feature of the Demaree method of swarm control is that the frames of eggs and brood, together with their nurse bees, are moved to the top of the hive above the queen excluder. The queen and all the flying bees are left in the lower brood box with an empty, clean comb or mite-free foundation. This alleviates overcrowding and prevents any swarming urge. After three to four weeks, as the brood above emerges and has destroyed or removed any queen cells, the two boxes can be merged, leaving a single brood chamber, possibly temporarily consisting of two boxes.

Stage 1: (Late winter) Check the box above the excluder to see how many bees, frames of honey and empty frames are there. If the weather is too cold to lift frames, look under the lid, remove the hive mat and look down into the frames.

Stage 2: (Late winter/early spring and before any queen cells) Exchanging frames across the queen excluder gives the queen more laying space. This increases the amount of brood and the number of adult bees, which will eventually be needed when the colony is split to prevent swarming. Queen cells can be destroyed, but only if the queen is present or if there are eggs in worker cells.

Stage 3: (spring) As soon as the colony is strong enough or there are queen cells, perform the Demaree method.

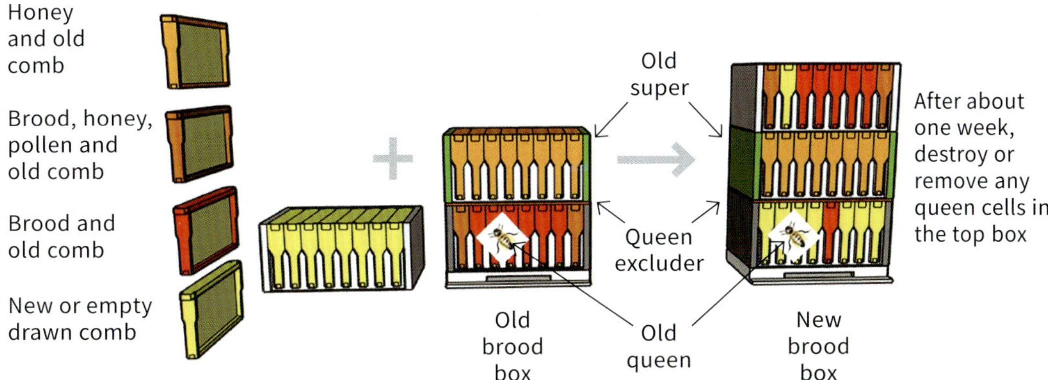

- **Destroy any queen cells, ensuring the presence of eggs in other cells before you do so.**
- **Move the old hive out of the way, and replace it with a new box and base. In that box, put the queen on a brood frame and fill the rest of the box with new frames with foundation or drawn frames. If the queen cannot be found, shake all the bees from the original box 1 into the new box 1.**
- **Replace the queen excluder.**
- **Replace the original box 2 on top of queen excluder (i.e. in the same position as the original box 2).**
- **Place the first box with brood but now without the queen, on top of the second box and add a frame with foundation to fill the space where the frame with the queen was removed.**
- **Six to eight days later, inspect the top box where the brood is present (i.e., box 3) and destroy or remove any queen cells.**

Stage 4: Inspect the brood box in four weeks to see if it is becoming too crowded. If necessary, add a second brood box by placing a box with empty (or drawn-out) frames under the queen excluder.

It is also possible to perform a somewhat modified Demarree method starting from a single box. In that case a new box is placed on the bottom board. A frame of brood is placed in the middle of that box with the queen. The rest of the bottom box is filled with empty frames (either drawn or with foundation). A queen excluder is placed on top of this box followed by the original box with a new replacement frame where the brood frame came from. As outlined above, check for queen cells six to eight days later and destroy any queen cells once the queen (or eggs) have been spotted in the bottom box.

What should I do when there are queen cells in my colony?

Problem: I can see what I believe are queen cells in my colony, but I do not know what to do. I am concerned that my colony will swarm.

Solution

Many beginners panic when they first see what they believe to be queen cells as they immediately think that the hive is about to swarm or has already swarmed. As a knee-jerk reaction, they often remove these cells immediately and close the hive again. This is a misguided strategy and has the potential to lead to swarming a short time later or leaving the hive without a queen and without the ability to make a new queen. Such a queenless hive unable to make a new queen will eventually die if left to its own devices. A much better strategy is to follow the steps below:

Establish that you see queen cells: While queen cells are very distinct from any other structure in the hive, it is essential to establish the difference between a queen cup without an egg and royal jelly inside a newly formed queen cell. To prove that, turn the frame upside down (the entrance to the queen cell is now pointing upwards) and look carefully for the presence of a white substance at the bottom of the cell (i.e., royal jelly). While the egg itself is hard to see, the amount of royal jelly present in a new queen cell is such that it is easy to spot. Queen cups are common in hives and do not represent any threat. They can also be removed without consequences, and some beekeepers remove the empty queen cups so that they do not have to recheck them at subsequent hive inspections. If the hive needs them, workers will make new ones, and if they are not required, removing them will make inspection faster and easier.

Establish what type of queen cells they are and at which stage they are: Decisions need to be made once it is established that there is a queen cell with royal jelly inside or a capped queen cell is present. The first step is to establish whether the queen cell is a supersedure cell or a swarm cell (see 'What is the difference between supersedure queen cells and swarm cells?'). If it is a supersedure cell, leave it alone and close the hive again. It does not matter at which stage the cell; the queen will eventually emerge and mate, and new eggs will be laid in due course. Ensuring the hive has not yet swarmed is essential if it is a swarm cell. Signs that the hive has swarmed include (i) there are fewer bees than the last time the hive was checked, (ii) the queen cell is capped, and (iii) the queen can't be spotted, or the queen can be found, but the marking is now gone (assuming the queen was marked). While it is theoretically possible that the workers have removed the markings from the back of the queen, this is quite rare.

If the hive has already swarmed: It is too late to prevent swarming, and nothing can be done about that. The focus should go towards preventing secondary swarming by checking for several queen cells in the entire hive (under the queen excluder if one was used). All but the biggest queen cell should be removed, preventing several queens from emerging. If there are still many bees, using some of these queen cells to start new colonies by performing a split might be possible. This has the advantage that there are several chances of getting a fertilized queen that is laying. Once that operation is done, it is also worthwhile carefully checking the environment (particularly ~2–7 metres (~5–20 feet) above ground, in tree branches) for the presence of a swarm. One can always be lucky that the swarm is still present and can be caught.

Allowing a hive to swarm is illegal in most countries, and swarm

Queen cells at the edge of brood frame are a sign that the colony is about to swarm. Queen cells in the centre of a frame indicate that the colony is preparing to supersede the old queen.

prevention is, therefore, compulsory. Even when it is not a legal requirement, swarming will result in losing many workers and honey. In addition, a natural swarm is less likely to succeed than a swarm that is managed within the apiary.

In any case, do not kill the queen cell unless you can see fresh eggs in the hive. Indeed, the presence of fresh eggs guarantees that if the queen is absent or infertile, the workers can make a new queen.

My colony is about to swarm; what should I do?

Problem: **There are signs that my colony is preparing to swarm, and I would like to prevent this from happening.**

Solution

Allowing a hive to swarm is illegal in most countries, and swarm prevention is, therefore, compulsory. Even when it is not a legal requirement, swarming will result in losing many workers and honey. In addition, a natural swarm is less likely to succeed than a swarm managed within the apiary.

Signs of swarming are the presence of queen cells at the early stages of development. These can be identified as queen cups into which an egg (and royal jelly) has been deposited. Cup cells without eggs and royal jelly do not represent a problem at this stage but should be monitored carefully if the queen decides to lay an egg in it and prepare to swarm. Once the queen cell is capped, it is likely that the hive has already swarmed, and it is too late to prevent swarming at that stage (unless the queen can be found, but this is unlikely).

Since swarming will occur once the queen cell is capped, it is possible to guess the available time to prevent swarming by estimating the likely date the egg has been laid (based on the developmental stage of the egg/larvae) and using Table 1 on p. 127. In most cases, however, time is short, and immediate action is generally required. The best way to proceed is to locate the queen and transfer her with half the bees/frames into a new hive. In the process, keep the frame with the uncapped queen cell in the hive without the queen. Select the best or largest queen cell and destroy all other queen cells. This prevents secondary swarms whereby a second queen emerging from a queen cell leaves the hive with some workers. This is a complete waste, as such a small secondary swarm headed by a virgin queen is generally unviable. The new hive with the one queen cell will continue to take care of that queen cell, eventually leading to a new queen that will mate and head this new hive. In effect, one forces the swarming by removing the queen and some workers and producing a new hive (see 'What should I do when there are queen cells in my colony?').

My colony is swarming; what should I do?

Problem: **A hive in my apiary is swarming (i.e., many bees are leaving the hive and settling on a nearby tree branch or other structure). I need to deal with this issue, but I do not know how.**

Solution

Observing a colony while it is swarming is an impressive sight as a mass of bees seems to stream out of the entrance to fill the sky with bees. Take note of where the bees are going and where they settle. There is no rush to collect the swarm as soon as it is settled, as the swarm will generally remain at the same spot from which they came for several days while the bees figure out the best place for their new hive. To maximise the chance of successfully collecting the swarm, leave the bees overnight where they are and collect them just before dusk the next day. Collecting them at dusk is best because they are now more ready to settle, having been hanging on a branch or other structure for over a day, and collecting them at dusk prevents them from immediately leaving the box in which they have been collected (they do not like to fly at night). By the time the sun comes up the following morning, the queen may even have laid some eggs. Once the queen lays eggs in the new box, the colony will be unlikely to abandon the hive. Another trick is adding a frame with some brood to the hive where the bees are collected, which implies that such a frame is available nearby.

How do I catch a swarm that I can see?

Problem: **I have noticed a swarm and would like to catch it.**

Solution

Suppose the swarm is on a structure that can easily be removed (for example, a tree branch that can be cut off). The easiest way to collect the swarm is to cut off the branch with one hand, holding it with the other, and gently deposit it into a box or a hive. Make sure there are some frames (preferably with empty comb) in the box or hive. Also, ensure some frames with foundation only are in the box. After swarming, the bees are primed to make comb and providing an opportunity for that will help prevent absconding. A good solution is to use two hive bodies, one with frames at the bottom and one without frames, but with the branch at the top. Leave the hive open so stray bees can enter. If the queen is in the box, stray bees will find her. You can generally observe workers fanning pheromones at the entrance to attract stray bees, a sure sign the queen is in the box, and all is well.

If the swarm has landed on a structure that can easily be picked up but not placed in the box, the easiest way is to shake the bees into a box with empty frames, including some frames with drawn comb and others only with foundation. Observe the bees at the entrance for fanning, indicating that the queen is in the hive and the swarm has been captured. If the queen is not in the box, the other bees will leave over time to rejoin the queen, possibly in the same spot which they came from. If this is the case, wait until the swarm is formed again and start over.

If the swarm lands on a structure that can't be shaken, brush the bees into the box. Constructing a funnel with a sheet of paper, cardboard or plastic may help guide the bulk of the bees into the box. A swarm of bees will act like a lumpy fluid when brushed. It does not matter how many bees are in the box, so long as the queen is captured. Indeed, as soon as the queen is in the box, the workers will fan at the entrance and attract all the stray bees, especially when it gets dark.

1. Swarms that are easily accessible can often be brushed directly into a small nuc hive.

2. Brushing a swarm off a tree trunk.

3. Once the queen is in the hive, the remainder of the swarm will soon follow.

4. A long pole with bucket attached can be used to collect swarms from high places.

How do I catch a swarm using a bait hive?

Problem: I want to trap a swarm using a bait hive rather than wait to see a swarm and catch it once it has settled on a structure.

Solution

As an alternative to catching a swarm once it has settled on a structure such as a tree branch or a fence post, to set up a trap for bee swarms. A trap typically consists of an eight- or ten frame beehive box with a bottom board and lid, with a few frames of old comb (avoid wax moth by freezing the frame beforehand) and a few empty frames with a starter strip of foundation. Add a commercial swarm lure or use a few drops of lemongrass essential oil soaked on a cotton bud. The smell of lemongrass mimics the scent of worker bee pheromones, and together with the smell of wax from the comb or foundation, it will attract a swarm to the empty hive.

Bait hives are often left in more difficult to access locations, higher off the ground. Placing lemongrass, old brood comb or a commercial attractant will increase the probability of catching a swarm.

The hive should typically be placed two to four metres above the ground in a shady but obvious area. The entrance should preferably point away from the midday sun (north in the Northern hemisphere, south in the Southern hemisphere). The distance from the apiary is unimportant and can be anywhere within a few hundred metres (~1000 feet). The hive should be checked regularly to see whether a swarm has taken residence and, if so, should be relocated to a convenient place soon after a colony has moved in (see 'How do I move a swarm?').

A bait hive is an empty hive left in a suitable location to attract a passing swarm. Often, a used brood comb or an attractant like lemongrass is left in the hive to attract scout bees.

How do I move a swarm?

Problem: **Now that I have caught the swarm, I need to transfer the swarm to my apiary or move the box to a new position in the apiary.**

Solution

A swarm is mainly composed of young bees, and they will reorient from wherever the new hive has been located. Hence, no special precautions are needed when moving a swarm, provided this is done before the bees can orient themselves by flying in and out of the hive. So, if the hive has been caught in the afternoon, the beekeeper has till the early morning of the next day to close off the hive entrance and move the hive. Generally, however, this will be done in the early evening as soon as most bees are in the box and the fanning at the entrance has stopped. Ensure the hive entrance is securely closed using either a closing mechanism (most nucs have this) or a ball of tissue/newspaper covered with strong adhesive tape (duct tape). While transporting bees, ensure they have plenty of ventilation and avoid exposure to hot temperatures (i.e., do not leave them in a car exposed to the sun).

Since a swarm needs to breathe while moving, closing the box may result in suffocation. Using a mesh bottom board is a good solution for this. Mesh bottom boards are inexpensive and easily obtainable from most beekeeping supplies. See image below.

Once the new location has been found, open the hive entrance, and let the bees explore their new environment and memorize the new location of the hive.

An inexpensive plastic nuc box can be used to move swarms. The lid needs to be secured and the entrance closed. Nuc boxes are usually provided with sufficient ventilation that bees can breathe even when the nuc is closed.

If a wooden hive is used to move a larger swarm, ventilation must be provided. This solution comes in kit form and consists of a mesh base, a mesh inner cover and a roof designed so that entrances on the side can be opened to allow air to circulate. The use of a mesh inner cover means the roof can safely be removed if the hive is transported long distances or the weather is particularly hot.

CHAPTER 12

Queens

Caged queens.

When should I re-queen my colony?

Problem: I need to know the signs that it is time to replace the existing queen with a new one (i.e., to re-queen the colony).

Solution

A queen can live for many years, and if she dies, she will be replaced by supersedure (see 'What is the difference between supersedure queen cells and swarm cells?'). During that process, the hive has detected that the queen is no longer present and will select several fresh eggs or lavae to start making a new queen. Thus, a colony can survive many years without having its queen replaced by the beekeeper. Nevertheless, many beekeepers replace the queen regularly, in some cases yearly or even more frequently. The signs to look for include the following:

The colony is aggressive: In many areas, it is illegal to keep aggressive bees; in any case, it is not pleasant for the beekeeper or neighbours. Aggressive hives can occur for several reasons, including that the original queen was aggressive, or that the colony replaced her without the beekeeper noticing. The newly produced queen has mated with drones from (generally feral) aggressive colonies in the area. Whatever the origin, the solution is the same: replace the aggressive queen with a gentle one. This needs to be done without delay, and since it is much faster to order a new queen than to produce one, this is often the pathway experienced beekeepers choose. Many queen breeders will supply queens bred explicitly for their gentleness.

A queen

Alternatively, it is also possible to kill the aggressive queen and merge the colony with another hive in the apiary (for example, a nuc many beekeepers keep for such situations).

The queen is a poor layer: From the onset or later in life, a queen sometimes produces only a small amount of brood. Under favourable conditions, especially in spring, one can expect several frames to be covered on both sides with brood. If this is not the case, it might be because the queen is not a good layer. A good comparison can be made with other colonies in the apiary subjected to the same environmental conditions.

There are signs of certain diseases: Different strains of bees have different capacities to deal with pests and diseases. If there are signs that the hive is affected by diseases or pests (other than AFB and *Varroa* mite), it might be possible to solve the issue by re-queening the hive. Indeed, providing the hive with a new vigorous queen will often overcome the problem, particularly if the strain used has some genetic resistance to the pest or disease encountered. Ask the queen breeder for a queen with these characteristics.

How can I find the queen?

Problem: I need to find the queen in my hive but have difficulties locating her.

Solution

Finding the queen requires practice and perseverance, and new beekeepers should practise the skill often, especially in the presence of experienced beekeepers. There is a certain degree of luck, but steps can be undertaken to maximize the chances of successfully locating the queen. Having two beekeepers look for the queen at the same time greatly increases the chances of spotting her on the comb.

Start by avoiding the excess use of smoke. Smoke results in the queen hiding in the corner of the hive, making it more difficult to locate her. Next, go directly to where the queen is most likely to be. The queen is much more likely to be found on a frame with some brood than on a frame with only honey. It is very unlikely that a queen would reside on a frame filled with capped honey or nectar as there is no laying space for her in that area of the hive. This means a super full of honey can largely be ignored when looking for the queen. Instead, focus on the middle of the hive where brood is likely to be found. Thus, after lifting the second frame from the edge of the hive out of the box to make space for easy removal of the other frames, go directly to the middle frames. Lift the frame out and quickly scan the perimeter of the frame to make sure that the queen is not on the edge. If she is on the edge, she will generally shy away from the light and move to the underside of the frame away from sunlight. Once the perimeter is scanned, systematically scan the entire frame, gently blowing on clumps of bees that may form to scatter them and make sure workers do not cover the queen.

In our experience, the queen is most likely to be spotted by the way she moves over the comb. The queen moves slower and in a more directional way than the workers, who tend to hurry in random directions. For beginners, queens and drones are easy to confuse. Focus on the drones' large eyes and rounded bottoms compared to the queen's slender figure and small eyes. By looking at the wings along the abdomen, confirm that it is indeed the queen. A queen will have wings finishing about halfway along the abdomen, while workers and drones will have wings along the entire length of the abdomen.

Of course, if the queen is marked, she will be much easier to spot, and beekeepers should consider either buying marked queens or marking the queen as soon as she is first found (see 'Should I mark my queen?').

Finding a queen in an aggressive hive can be challenging and is best done by at least two experienced beekeepers. There are some specific techniques to make this easier, including the following:

- **If the hive has several boxes, place a queen excluder between each box and wait a week. By that time, only one box will have open brood (i.e., visible larvae). That box contains the queen, immediately reducing the search area to a single box.**
- **Move the box with the open brood away (say 10 metres (10**

yards) or so) and allow the foragers to return to the main hive. This will limit the number of aggressive bees in the box to be inspected.

- Work methodically, always leaving a gap in the box between the frames that have been inspected and those that still need to be done. This prevents the queen from getting back on a previously checked frame.
- Focus on frames with brood, as it is much more likely that the queen will be on a frame with brood than a frame without brood.

How can I tell whether the queen is present in the hive?

Problem: **A queen can be challenging to find, particularly in a large, aggressive hive. I want to know with some degree of certainty that the queen is present without locating her.**

Solution

The fact that the queen can't be located does not necessarily mean the queen is absent. Indeed, queens can easily be missed, and there is a limit to a beekeeper's patience and the amount of time one can reasonably look for the queen in a hive. As a rule of thumb, we would look at every frame twice. If, after looking at every frame twice, the queen can't be found, it is best to close the hive and start again the next day when the bees have settled. This is especially true when the hive is aggressive, and the bees try to sting the beekeeper as he looks for the queen. To facilitate locating the queen the next time, place a queen excluder between each box so that the queen is confined to the one box and that box can be identified a week later because it will be the only box with eggs and/or open brood.

While it is not possible to be sure that the queen is present without

Eggs and young larvae in brood comb. This is a sign that the queen was present up to a day ago.

spotting her, the presence of fresh eggs suggests that the queen was there less than 3½ days ago (See Table 1. Also, see 'How can I predict what will happen in my hive several weeks in advance?'). Experienced beekeepers can distinguish one-day-old eggs from older eggs by the position in the cell (upright at the start and flat at the bottom of the cell later), so they can narrow that down to less than 3½ days. If the hive has not been opened in the previous 3½ days (i.e., no chance of the beekeeper accidentally killing the queen), and eggs can be seen, one can be very confident that the queen is inside the hive somewhere. While this does not provide certainty, this high probability is generally sufficient to decide on the best course of action.

However, the reverse is not valid as a queen may be present without eggs present in the hive. There are two reasons for this.

- **During very cold periods, the queen generally stops laying. This occurs when the hive is clustering and unable to maintain brood at 37 °C (100 °F).**
- **Since it takes 21 days after the eggs are laid for workers to emerge and it takes around four weeks for the queen to start**

laying, it is possible that a new queen was made and that she had not yet begun to lay. Checking the hive a week later will resolve the issue, as by that time, fresh eggs should be present if a new queen was made. Therefore, the absence of eggs one week later indicates the lack of a queen.

What strain of queen should I buy?

Problem: **I plan to requeen my colony and want to know which strain I should choose.**

Solution

The genetic background of the queen and the drones she has mated with has a major impact on the bee colony. Indeed, all the workers in the colony are either sisters or half-sisters and hence will have similar characteristics as the queen with some impact of the drones which mated the queen. Hence, selecting a strain of queen that satisfies your needs is critically important to not only honey production, but also gentleness, disease resistance, propensity to swarm and abscond, as well as appearance and climatic adaptation, such as early build-up of bees in spring.

When selecting a honey bee queen, there are several factors to consider based on your location-specific needs and beekeeping goals. Considerations to help you choose the right strain include the following:

- Different strains of honey bees have adaptations to various climates. Ensure the strain suits the local environment where you'll be keeping the bees. Thus, ordering queens produced in vastly different climate zones should be avoided.
- Some strains are known for certain behaviours, such as gentleness, productivity or disease resistance. Research the behavioural traits of different strains to find one that matches your preferences.
- Certain strains may be more prolific honey producers than

others if you are primarily interested in honey production.
- Some strains exhibit greater resistance to common diseases and pests, which can reduce the need for interventions and treatments.
- Depending on your location, some strains may be more readily available from local breeders or suppliers.

Common strains of honey bee queens include the following:

- Italian bees are favoured among beekeepers worldwide for their gentle nature, high honey production and early spring buildup. Their adaptability to various environments and strong foraging abilities make them well-suited for commercial and hobbyist beekeeping operations.
- Carniolan bees are favoured by beekeepers for their gentleness, rapid spring buildup, cold hardiness and some level of resistance to pests like *Varroa* mites. These traits make them well-suited for various climates and appealing to beekeepers looking for productive and manageable honey bee colonies.
- Caucasian bees are favoured among beekeepers who prioritize gentleness, cold hardiness and resistance to certain pests. Their adaptability to cooler climates and their propensity for propolis production make them valuable in regions where these traits are advantageous.
- Russian bees have gained popularity among beekeepers, particularly those interested in sustainable beekeeping practices and reducing reliance on chemical treatments for pest management. Their natural resistance to *Varroa* mites and adaptability to colder climates make them attractive for beekeepers in various regions.
- Buckfast bees are valued for their gentleness, productivity, disease resistance and adaptability. They have been selectively bred to combine desirable traits from various honey bee subspecies. This makes them popular among

1. Italian, Caucasian and Carniolan bees are popular strains to keep. Italians are a yellowish-straw colour.

2. Caucasians are dark.

3. Carniolans are a musky-grey.

beekeepers looking for reliable and productive bees for hobbyist and commercial beekeeping operations.

These are the most common strains purchased. Depending on the country in which you live, many other strains, or hybrids, are produced locally that may be better acclimatized to your area. Ask local beekeepers or perform a web search to determine if any specialty queen-rearers are near you.

Before buying a queen bee, ask local beekeepers which strains they find best in your area and why. Also, talk with several queen-rearers and ask what the characteristics of their bees are. Each rearer produces bees with different advantages and disadvantages, such as disease resistance, honey production, gentleness, reduced swarming and better overwintering. Ensure you understand the characteristics of the queen you plan to buy and if she will produce a colony with the required traits.

How should I re-queen my colony?

Problem: I have decided to re-queen my colony and do not know how.

Solution

A colony can only have one queen, so removing the old one before introducing a new one is essential. Failing to do this will result in the queens fighting, with the possibility that the newly introduced queen will be killed. The first step is to locate the old queen (see 'How can I find the queen?'). The queen needs to be killed, which can be achieved quickly by squashing her with the hive tool or removing her into a queen cage or catcher.

Once the hive is queen-less, the new queen can be introduced (see 'How should I introduce a new queen into a queen-less colony?'). If the reason for requeening the hive is that the hive is aggressive, we would advise immediately implementing the procedure explained in 'What should I do if the colony is killing my newly introduced queen?' Indeed, there is a high degree of probability that the queen will be killed if she is introduced directly. Introduction via an intermediate step (i.e., first into a nuc containing gentler bees, followed by merging) reduces the risk of rejecting the queen.

See 'When should I re-queen my colony?' and 'How should I introduce a new queen into a queen-less colony?'

How should I introduce a new queen into a queen-less colony?

Problem: I have a queen-less colony and want to introduce a new queen.

Solution

Purchased queens arrive in the mail in a specially designed plastic or wooden cage with several nurse bees. These nurse bees care for the queen by feeding and grooming her. These are essential as the queen is unable to feed herself. The cage generally has a mesh-covered area where the queen can interact with the workers in the hive and hide from external bees should she feel the need to be temporarily separated from these workers, for example, if they display aggressive behaviour towards the new queen. The cage is generally closed by a sugar plug, sometimes covered by a plastic cap during shipment or storage, which must be removed before placing the cage in the queen-less hive. This plug acts as a timer for the queen's release, as it will take some time (usually a few days) for the workers from the hive to chew through the sugar plug, at which stage the workers will have accepted the queen and she can walk out and join the queen-less hive.

To introduce the queen to the queen-less hive, the queen-less hive is first smoked extensively to mask any smells from the old queen that may have remained and any from the newly introduced queen. A frame with some brood is selected, and the queen cage is gently pressed into the comb close to where the brood is. Be careful to only slightly push the cage into the comb so as not to obstruct the sugar-plugged opening, and with the

INTRODUCING A NEW QUEEN

1. A cage used to ship a queen. Cages may be made of wood or plastic.

2. A plastic cage holding a queen is ready to be placed in a queen-less hive. Also, in the neck of the cage, white honey-sugar candy can be seen, used to feed the queen during captivity, and to delay her escape from the cage while she eats her way out. A queen cage is pressed into comb, near the brood area, where she will emerge in two or three days, allowing the workers to become used to her scent.

3. A cage holding a queen pressed into brood comb deeper in the hive. Nurse bees will be near the brood and look after the queen, providing her with food and water. In two or three days, the queen will emerge and the worker bees will have accepted her.

4. When a colony realises a queen is in a cage, they try to reach the queen, covering the cage with worker bees.

opening facing 45° upwards. This is necessary so that if one of the nurse bees accompanying the queen dies, she will fall to the bottom of the cage and not block the queen's escape route. Replace the frame in the hive and close the hive. Inspect the hive in a few days to see if the queen is out of the cage. If she is still in the cage, manually remove the plug and allow the queen to escape. Check again a week later that fresh eggs and larvae are present, a sure sign that the queen is laying. Indeed, after a week, only newly laid eggs will be at the open brood stage (i.e., eggs and larvae), while older eggs will have matured to capped brood.

If there is no brood in the hive when introducing the queen, collect a frame with brood from another hive, shake off all the bees and introduce it into the queen-less hive a few days before introducing the new queen. This will achieve two goals: (i) it will ensure that there are recently emerged workers in the hive that will take care of the queen and her offspring, and (ii) it will significantly increase the chances of the queen being accepted as the colony now thinks that the new queen has already laid eggs and is more likely to give her a chance.

What should I do if the colony kills my newly introduced queen?

Problem: My queen-less hive has killed the newly introduced queen, and I want to prevent this from happening again when reintroducing a new queen.

Solution

Sometimes, a new queen gets rejected, especially with a large, aggressive colony. Experienced beekeepers can detect aggressive behaviour towards a new queen at the time of introduction, but the signs are pretty subtle. Several precautions can be applied to prevent the queen from being killed.

Make sure the queen has an area in the cage in which she is introduced where she can hide from the colony. While most modern cages have this feature, if it is not present, use tape to block off a small portion of the queen's cage so she can seek protection.

Prepare a small queen-less colony with plenty of brood using a gentle colony in a five frame nuc. This can be done from any hive in your apiary. Introduce the queen in this nuc as described above. Ensure the queen is laying by checking for open brood a week after the queen is released. Use newspaper to merge the nuc with the large, aggressive queen-less hive. Check ten days after the merger (two to three days for the hives to merge and a week for the old eggs to be matured to capped brood) for signs of open brood (i.e., eggs and larvae). Indeed, in the unlikely event that the queen gets killed by the aggressive hive, it will happen very soon after the merger, so the presence of open brood is a sure sign that the queen is laying. If the new queen freshly lays eggs, the queen-less hive will make its queen from these fresh eggs.

Use plenty of smoke when introducing the queen or merging the hives to mask the new smells when these operations occur.

My new queen arrived during a cold, rainy period; how can I introduce her to my colony?

Problem: A newly purchased queen generally arrives in the mail and could be delivered when it is too cold and wet to open the hive for the time required to introduce the queen to the colony. I wonder what I should do under these circumstances.

Solution

The caring beekeeper will always aim to open the hives when the weather is favourable. Ideally, this would be on a windless, sunny day with a temperature of at least 18 °C (65 °F) and no more than 37 °C (100°F). These conditions are not always available, and it is possible to open the hive in less optimal conditions. However, there are severe risks when inspecting the hive during cold/wet periods, including the heightened risk of chalkbrood and nosema developing (see Chapter 19 on 'Other pests and diseases'). If the hive must be opened during these periods, aim for opening the hive as quickly as possible during mid-morning, and take out as few frames as possible, leaving the hive as much time as possible to warm up again before the colder night temperatures arrive.

It is, however, often better to wait a few days and open the hive under better circumstances (check the weather forecast). The queen can usually remain in her cage for several weeks provided that she has mated, and the sugar plug is intact, that nurse bees are supporting her and that she gets a drop of water every few days. She also needs to be stored in a cool, dark place. Do not be tempted to give her sugar water, as that may well stick to the queen, which will likely kill her. Instead, provide a drop of water suspended by surface tension in the cage mesh and let the nurse bees collect it and use it to dissolve some of the sugar plug.

What is the difference between swarm, queen cells and emergency cells?

Problem: **I need to know whether my hive is about to swarm (presence of swarm cells) or whether the queen is being replaced because she died (emergency).**

Solution

It is essential to be able to tell the difference between swarm cells and emergency cells because the beekeeper needs to perform different procedures in each case. If there are uncapped swarm cells, the beekeeper must act urgently to prevent swarming. In contrast, the presence of emergency cells does not require urgent treatment, as the hive is likely to eventually continue with a new queen, as before.

It is essential to realise what the bees are doing to tell the difference. In the case of swarm cells, the bees are trying to reproduce the colony. This is a planned process whereby the workers make queen cups at a convenient location and wait for the queen to lay an egg into these queen cup cells. Making queen cells for swarming is a planned process; the cup cells are invariably made vertically, generally at the bottom of the frame or in a cavity in the comb.

On the other hand, emergency cells result from a response to an emergency, namely that the queen is either incapacitated or dead. Hence, there is no time to make a queen cup at a convenient location and wait for the queen to lay an egg in it. Therefore, emergency cells are typically located wherever a fresh egg is present, generally in the middle of the comb. The emergency cell also does not hang properly, as is the case for a swarm cell, but is protruding and curved out of an ordinary worker cell within the comb.

Emergency cells should never be removed as they generally represent the last viable option for the hive to make a queen. Typically, it is best to leave one queen cell in a hive if no fresh eggs can be found. If a queen cell is found to be capped it is also essential not to remove this queen cell as it is either an emergency cell (and therefore the last chance of the hive to make a new queen) or it is a swarm cell. Still, in that case, the swarm has most likely already left (unless the queen can be spotted), as the hive swarms as soon as the swarming queen cells are capped. Another sure sign that the hive has swarmed is that hive appears to have fewer bees than during previous inspections. If the swarm has already left, the old queen has also gone; therefore, the hive needs the queen cell to replace her. So, do not remove the swarm cell in that case.

Should I mark my queen?

Problem: When I inspect the brood box, my queen is difficult to find. I am considering marking her to make her more visible. Marking the queen will also allow me to confirm whether she has been replaced.

Solution

Most new beekeepers have difficulty finding a queen, and this process is greatly facilitated by marking the queen with a special marker pen. The coloured dot stands out amongst the many workers and drones and, therefore, allows for easy identification of the queen. Another advantage of having marked queens is that if the queen is replaced by supersedure or swarming, she will no longer be marked. Therefore, these events can be identified by the beekeeper. Consequently, we believe it is good practice to mark the queen. A queen-rearer (i.e., a beekeeper focused on rearing queens) marks most purchased queens using a colour code on the back of the queen corresponding to the year in which the queen was produced. High-value breeding queens are often marked by gluing a small, numbered disk to the queen's thorax or torso.

Queens are marked with one of five colours to indicate the year the queen was reared. See Table 3 below:

Colour	Year reared
White	Years ending in 1 or 6
Yellow	Years ending in 2 or 7
Red	Years ending in 3 or 8
Green	Years ending in 4 or 9
Blue	Years ending in 5 or 0

Table 3

For example, a queen reared in 2025 would be marked with a blue pen, and a queen reared in 2026 with a white pen. A good mnemonic for remembering the colours is 'Wow, You Raise Great Bees' (White, Yellow, Red, Green, Blue).

Most beekeepers use a Posca pen since they are readily available and non-toxic. The pen marks the queen with the year she was reared, allowing identification of older queens that need to be replaced. Some beekeepers hold the queen between the thumb and index finger to mark a queen, while others prefer to place the queen in a cage or a plunger and tube. All methods are practical, and the chosen method will depend on the beekeeper's preference.

Using a non-toxic, water-based POSCA pen and cage to mark a queen.

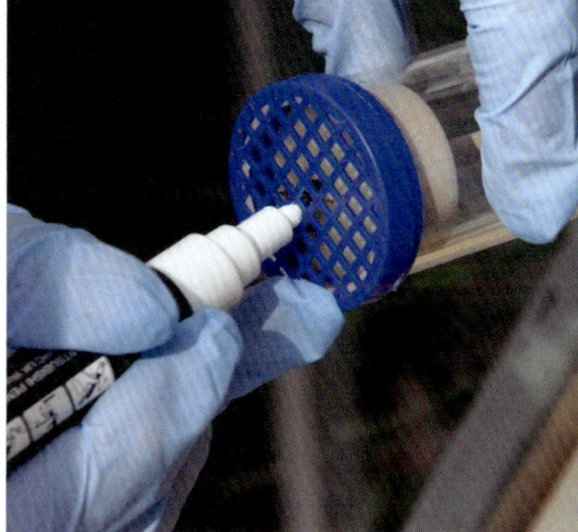

A plunger and tube are convenient ways to mark a queen.

What should I do when my queen is getting through the excluder?

Problem: I found brood above the queen excluder, suggesting that the queen has gotten into the honey supers.

Solution

There could be several reasons why the queen is above the queen excluder, including a gap in the excluder or area surrounding the excluder, inadvertently letting the queen above the excluder during the previous inspection, or the queen is so small that she can pass through. The result is that she has been laying eggs in the honey supers.

The first step is to find the queen; we now know that she is likely on top of the excluder. Although if there is a gap in the excluder or if she is so tiny that she can pass through the excluder there is a slight chance that she might have returned downstairs. Nevertheless, start to check the honey super for the queen systematically. When she is found, bring her back below the excluder and check for gaps that might have allowed her through. The brood can be transferred below the excluder, but this is optional as the brood will have emerged in a maximum of three weeks. It should be noted that drones cannot leave through the excluder, so any drone brood should be relocated below the excluder, or there will be a layer of dead drones on top of the excluder at the next inspection. Check again a few weeks later to make sure that the queen stays below the queen excluder.

Eggs or larvae in the super often mean the excluder is broken, which needs to be repaired or replaced.

CHAPTER 13

Queen Rearing

A newborn bee.

Can I rear queens?

Problem: **I want to rear queens, but I must figure out where to start.**

Solution

There are several ways of approaching queen rearing. In its simplest form, it is possible to rear queens just by performing splits (see 'How do I prevent swarming and produce more hives?' and 'How do I split a hive?'). Suppose one intends to expand the number of hives rapidly. In that case, it is often possible to do two or three splits each year, but only with strong colonies and during periods of high nectar flows (depending on your area, this is generally in spring but could extend to summer, unless your summers are hot and dry). It is generally not a good idea to perform splits in autumn as this does not leave much time for the hive to increase in size and produce sufficient stores. The hive/colony will most likely need to be fed.

It is possible to rear queens using grafting from a colony with a different genetic background from the colony one wants to split, but it requires

Queen pupae maturing after grafting.

A QUEEN EMERGES FROM HER CELL

1. A queen cell containing a queen. 2. The virgin queen chews away the top of the capped queen cell, unaided. 3. The virgin queen emerges. Her abdomen can be seen as she scuttles away at the bottom-right of the capped cell. 4. The young queen stretches her legs, alone, at the bottom of the cell.

great care, precise timing, and technical skills best learned from an experienced queen breeder. The principle is to produce a queen-less hive without the ability to make a new queen because there are no fresh eggs or young larvae. This may require removing all queen cells within a week to ten days of making the hive queen-less thus ensuring the colony cannot rear its own queen. If left to its own devices, such a hive will perish as there is no queen and no ability to make a queen. Placing eggs or young larvae in a queen-less hive within the first week after the hive is made queen-less will result in the bees eagerly making many new queens from the newly introduced larvae. The freshly introduced larvae can come from a frame transferred from another hive, from grafting or from artificial cells in which the queen from another hive has laid eggs.

What are the advantages and disadvantages of different methods of rearing queens?

Problem: **Rearing queens demands a lot of effort and is uncertain, but it is also rewarding and has distinct advantages.**

Solution

The disadvantage of producing one's own queens is that it takes time and effort and can quickly go wrong because of the many steps involved, even if one is experienced. Indeed, there are several steps over which the beekeeper generally has little control. One of the most important is the queen's mating, which occurs several kilometres from the hive and is notoriously difficult to influence. The most important aspect of being unable to control which drones the queen will mate with is that she will likely mate with some drones from feral colonies. Typically, feral colonies have a more aggressive temperament, giving them an evolutionary fitness advantage. Hence, home-reared queens tend to have more aggressive offspring. Professional queen breeders go to great lengths (flooding the area with known drones, instrumental insemination and timing of queen release for mating) to at least partially control the drones the queen mates with and most likely include honey production and gentleness as two of the selection criteria. These methods are very time-consuming and expensive and are out of reach for the average amateur beekeeper. Hence, queens purchased from a reputable queen-rearer generally lead to gentler queens. This is particularly important if the area you are keeping your bees in is home to Africanized bees (i.e., 'killer bees') in North and South America (see 'What are Africanized bees, and how do I prevent them in my hive?').

On the other hand, rearing your own queens is cheaper and more rewarding than purchasing them every year or two. We also believe that this makes you a more knowledgeable beekeeper, forcing you to learn more about bee behaviour, and allows you to control your beekeeping activities better. In addition, rearing your own queens will, over time, produce bee strains that are better adapted to the environment where you keep your bees. Another advantage is that it tends to preserve the genetic diversity of the honey bees and increases the resilience of colonies in your area.

Rearing your queens and buying new queens do not need to be mutually exclusive, and it is okay to supplement your queen rearing with the occasional purchase of a replacement queen, particularly if the colonies you keep become aggressive (see 'What should I do if my bees are aggressive?').

Grafting remains the most widely used method to rear queens. See image on p. 180 / 181. for mature grafted pupae in capped plastic queen cups.

How do I split a hive?

Problem: **I want to split my hive to make a second colony and wonder how this is done.**

Solution

Splitting a Langstroth hive manages bee populations and creates new colonies, including colonies that may soon swarm. The steps to be taken are:

1. Inspect the hive and look for the queen:

- Smoke the entrance to calm the bees, if needed. However, smoking the bees may make finding the queen harder.
- Make sure the brood chamber includes two hive boxes.
- Open the hive and gently inspect the combs. Look for queen cells and brood patterns. Ideally, you want to split when the hive prepares to swarm or has a strong queen.
- If the hive only has one brood box or has a weak colony, merge with a second strong colony.
- Check that eggs and young larvae no more than one day old are in both boxes.

2. Split the hive if you can find the queen:

- Determine which brood box the queen is in (see 'How do I find the queen?').
- Place a divider board between the two brood boxes.
- Make sure the queen-less brood box contains either eggs or larvae no more than than one day old.
- The queen-less brood box may already contain queen cells, which may have been the reason for the split of the hive.
- The divider board needs to include an entrance at the rear of the hive, while the entrance to the brood box with the old queen faces the front.
- Close the hive by replacing the lid.

- Check the queen-less brood box the next day or the day after to confirm that the workers have started building queen cells.
- Monitor both hives for a few weeks.
- Check for egg-laying and overall activity to ensure the colonies function well.
- Ensure they have enough food and space to grow.
- Alternatively, place the brood box with the queen on a new bottom board and with a new lid at a different location. Make sure to add plenty of bees and stores.
- The hive at the old location will make a new queen and benefit from plenty of foragers bringing in supplies to create a healthy queen.
- The foragers will return to the old location; hence, do not expect to see a lot of activity at the entrance of the hive with the queen at the new location.

3. Split the hive if you can't find the queen.

- If, despite looking carefully, you cannot find the queen, it is still possible to split the hive.
- Proceed as above, but since you do not know where the queen is, it is essential to have fresh eggs, larvae and plenty of stores and workers in both brood boxes.
- Check both boxes to see which one has the queen. The box that starts making a queen cell does not have a queen.
- Leave the box with the starting queen cell in the original location, and move the box with the old queen to the new location to maximize resources coming into the hive that is making a new queen.
- The foragers will return to the old location; hence, do not expect to see a lot of activity at the hive's entrance with the queen at the new location.

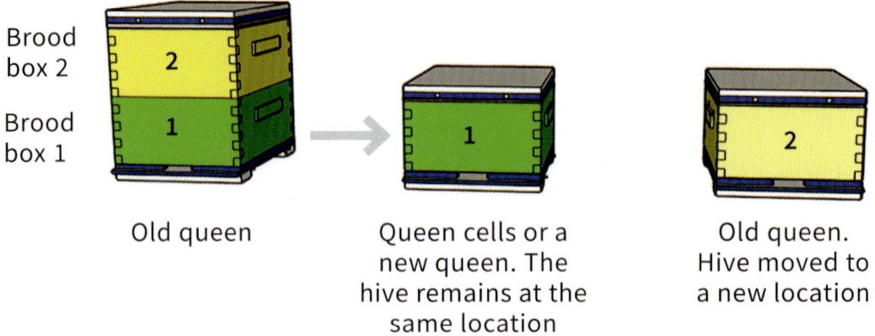

SPLITTING A HIVE

Two brood boxes and two sets of lids and bases are needed to split a hive. Separate the brood boxes and give each hive its base and lid. Different steps are now taken depending on whether you have found the queen. See the text for an explanation.

Before and after splitting a hive, it is good practice to feed all hives with pollen and syrup, ensuring the colonies are well-nourished and able to support a growing population. Splitting a hive is a great way to manage bee populations and prevent swarming, and with practice, you'll get more comfortable with the process.

How do I manage colonies to rear queens?

Problem: The production of queens requires several hives to be prepared in advance. A hive must be made queen-less to accept grafted eggs and produce queens. Later, several smaller hives, called nucs, must be available to accept emerging queens and allow them to mate.

Solution

1. A queen-less hive with enough young worker bees to produce royal jelly and form the queen cell is needed to create a queen. These hives are called 'cell builder hives' and are obtained by removing the queen and, after a week, checking the hive for queen cells produced from

fresh eggs. These queen cells must be destroyed, so the hive has no queen or way of making a new queen (i.e., all the brood is now too old to be used to form a queen). Since fresh eggs take three weeks to emerge as workers, such a hive will still have plenty of young workers for the next several weeks and is now ready to accept any grafted larvae and transform them into a new queen.

2. Young (i.e., less than 24 hours since hatching) larvae from a second hive need to be grafted into plastic queen cups and placed vertically in the queen-less hive, as this is the orientation of a queen cell. The colony in this hive from which the larvae are grafted usually has superior traits or genetics that you would like to replicate. Superior traits include greater honey production, gentleness or disease resistance.
3. About ten days after transferring the fresh larvae to the queen-less hive, it will be essential to have one queen-less colony available to accept each of the capped queen cells when they are ready to be transferred from the cell builder hive. These new colonies need not be large; nucs or even small mating colonies are sufficient. However, small colonies are often challenging to keep as they are more prone to overheating, starving or absconding. Thus, starting beekeepers are generally better off using five frame nucs rather than smaller hives.

What is the timeline for queen-rearing?

Problem: A bee's life cycle timeline is tightly controlled and predictable, used to our advantage when rearing queens. As such, it is possible to anticipate what will happen at each stage over the next few days or weeks.

Solution

Queen rearing is highly dependent on knowing and exploiting bees' life cycle and behaviour. Fortunately, this process is highly regulated within the colony and predictable, with often only minor variations in the timing (say +/- a few days at most). See Table 1 on p. 127.

Within a few hours of the hive becoming queen-less, through the queen dying or because of beekeeper intervention, the queen pheromone levels will have diminished to such an extent that workers will become aware the queen is no longer there. This will result in a rapid process of selecting young larvae that will be raised as queens in no more than 48 hours. Typically, six to twelve larvae are chosen this way. The workers feed these larvae copious amounts of royal jelly and start making queen cells around the larvae. Suppose the hive has been queen-less for some time and could not produce a queen (for example, because the beekeeper has removed all the newly produced queen cells). In that case, the hive is keen to make a new queen and will attempt to transform any newly introduced egg into a new queen.

From Table 1, it can be inferred that the selected egg will hatch within

Table 4: Sequence of events when preparing hives for grafting.

SUNDAY	MONDAY	TUESDAY	WEDNESDAY	THURSDAY	FRIDAY	SATURDAY
1 Remove Queen from Hive 1	2	3	4	5	6	7 Eggs laid in Hive 2
8	9 Check Hive 1 for queen cells. Destroy any queen cells	10 Graft young larvae from Hive 2 and place in Hive 1	11	12	13	14
15	16	17	18	19	20 Remove each capped queen cell from Hive 1 and place in its own nuc	21
22	23 Eggs laid on day 7 hatch in nuc colony	24	25	26	27	28 Check each nuc that the queen has emerged and has started to lay eggs

1

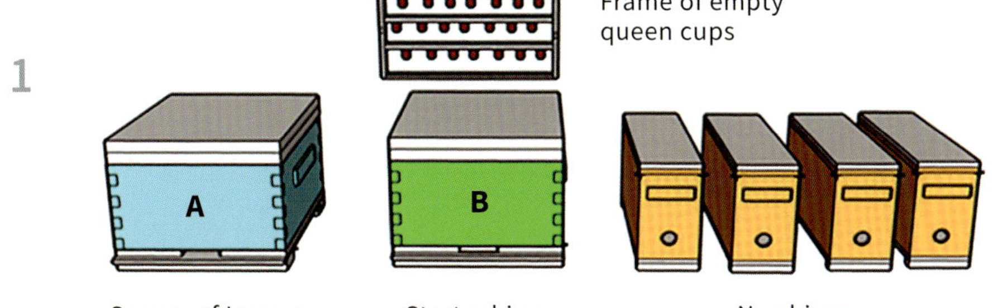

2

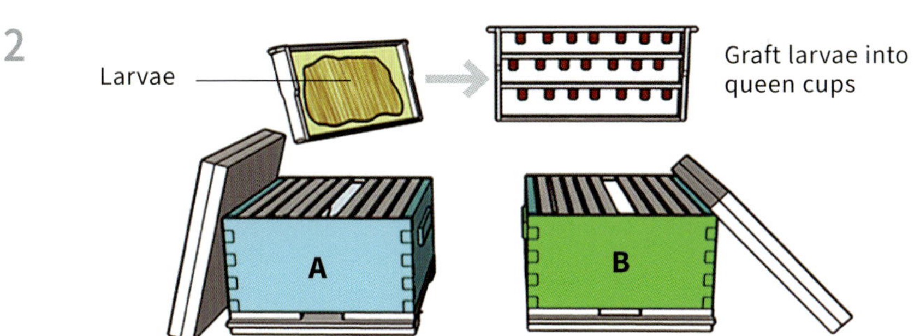

3

Starter hive with grafted larvae
left for ten days after grafting

4

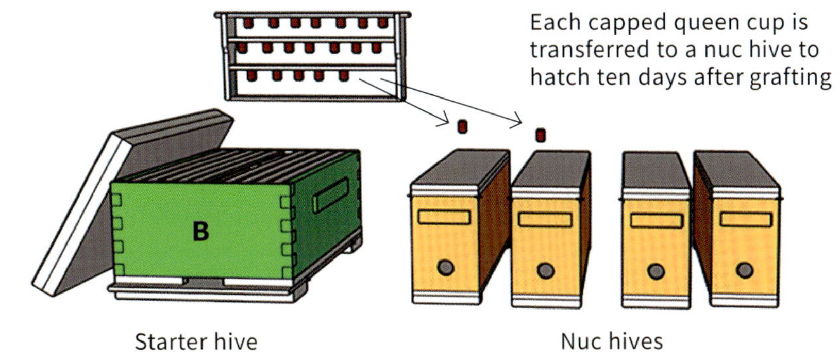

3.5 days from the eggs being laid and that the queen cell will be capped after eight days within a window of one day (i.e., between seven and nine days). The queen will emerge between fourteen and eighteen days after laying the egg. This is critical, as any emerging queen will likely kill all other queens while still trapped in their cell. Thus, a beekeeper wanting to produce more than one queen within the same hive needs to remove the queen cells just after they are capped and well before they emerge (typically about ten days after grafting, which corresponds to a maximum of 11 days after the egg was laid since the larva was no more than two days old when grafting took place).

After emerging, the queen will remain in the hive for one to two weeks, maturing and waiting for suitable weather for her mating flight. She will start laying a few days after returning from her mating flight, typically between 23 and 32 days after the egg she was generated from was laid.

How do I graft larvae to rear queens?

Problem: **I have a colony of gentle, productive bees from which I want to rear many queens.**

Solution

Grafting transfers young larvae from a colony that has traits or genetics you want to propagate, such as honey production, gentleness or disease resistance, to a queen-less hive to produce queens. Grafting is a skill that requires patience and techniques best learned from an experienced beekeeper. The first step is identifying larvae of suitable age (less than 36 hours; generally, the smallest larvae on the frame). Unless you have excellent eyesight, a magnifying glass or, better, a 10x magnifying light is used. Preventing the larvae from drying out is crucial, so speed is essential after removing the frame from the hive. A damp towel covering the brood can help prevent the larvae from drying out in the frame.

Using a grafting tool (several types are available from beekeeping supply shops or reputable online suppliers), the larva and surrounding royal jelly are scooped out of the cell and transferred to a wax-coated plastic queen cup. These cups have previously been attached to a frame, so they hang vertically when placed into the hive. Generally, ten–twenty queen cups are attached to one frame. The frame with the grafted larvae in the plastic queen cups is placed into a strong, queen-less hive that contains brood but no fresh eggs or young larvae.(see 'How do I manage colonies to rear queens?'). This is essential as the production of queens requires recently emerged worker bees that will produce royal jelly and form the queen cell around the queen cup. The grafted frame is left in this hive for ten days; fully formed capped queen cells should be present at this stage. This colony should be fed syrup to ensure the workers are well-nourished and have plenty of royal jelly to feed the introduced queen larvae.

These queen cells need to be transferred to new individual hives (one queen cell per hive) so that over the next week, the queen can emerge. Since there is now only one queen cell in the hive, the emerging queen will not kill the other queens. As an additional precaution, in case the beekeeper does not get the timing quite right, there are plastic cages that fit over each queen cell and trap the queen, preventing her from reaching the other queen cells and killing these queens by stinging through the queen cell.

Since grafting is technically challenging, Nicot (French) and Jenter (German) make similar commercial systems that force the queen to lay eggs in detachable plastic wax-coated cups that can be transferred between hives. Hence, grafting is not required with this system. The bee community is split as to whether this system is easier than grafting, and the authors are aware of several imitation products that are poorly designed and do not perform as well as the original. The Nicot and Jenter methods are not discussed in this book.

Equipment for rearing queens is available from beekeeping stores and online. Equipment needed to graft includes:

- Grafting tool
- Plastic queen cups
- Head-mounted lamp if a light is needed
- Glasses or magnifying glass if needed
- A strong, queen-less colony to hold the grafted larvae while they develop into queens
- Nuc box to place queen cells a day or two before the new queen emerges
- Queen cages to hold mated or unmated queens
- Queen marker pen to mark the queen
- Cage to hold queen while she is being marked

A grafting tool is used to remove larvae from brood comb and place it in a plastic queen cup.

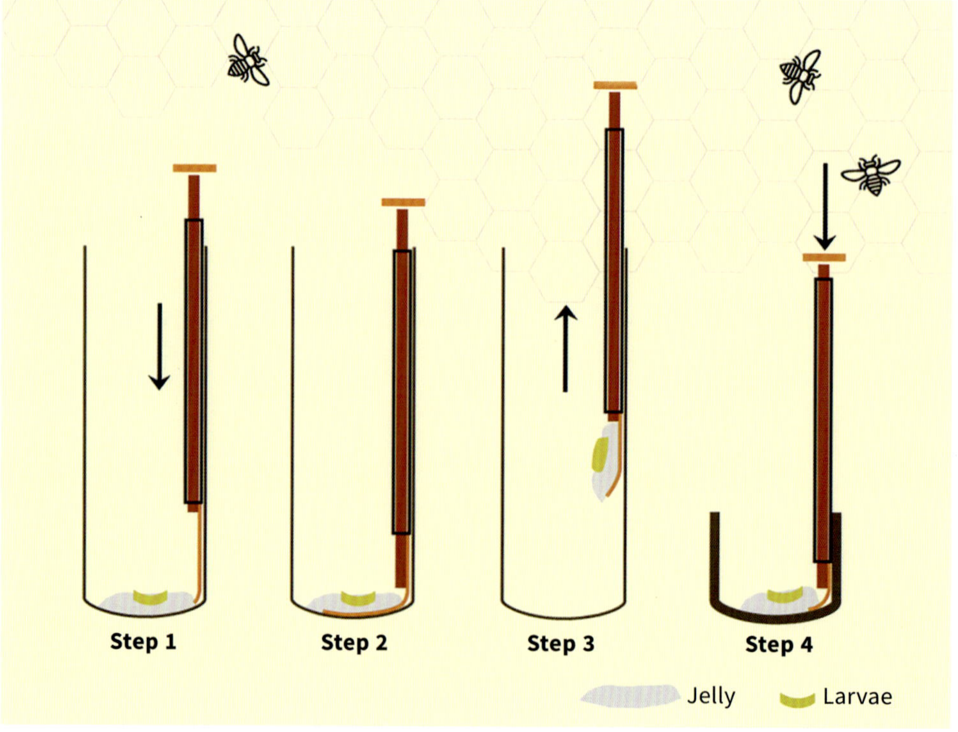

Eggs, larvae and pupae. Only larvae one to two days old can be grafted. Larvae suitable for grafting are in the centre-left cells.

Place young larvae into a plastic queen cell, making sure the walls of the cell are covered either in royal jelly or sugar syrup.

A frame of plastic queen cells ready for grafting.

How can I store queens until they are needed?

Problem: **I have produced some mated queens, but I am not ready to use them and would like to store them before shipping or transferring them to a permanent colony.**

Solution

Once a queen has been transferred to a mating colony or nuc, has mated and is now laying eggs, this queen is ready to head her colony. Of course, it is possible to leave her in the mating colony and to grow this colony into an entire hive. However, in some cases, the mating nuc will be used for another round of queens being produced, and the queen must sometimes be stored for prolonged periods. Queens can be caged individually, and

Below: Different designs of queen holders used to store queens.

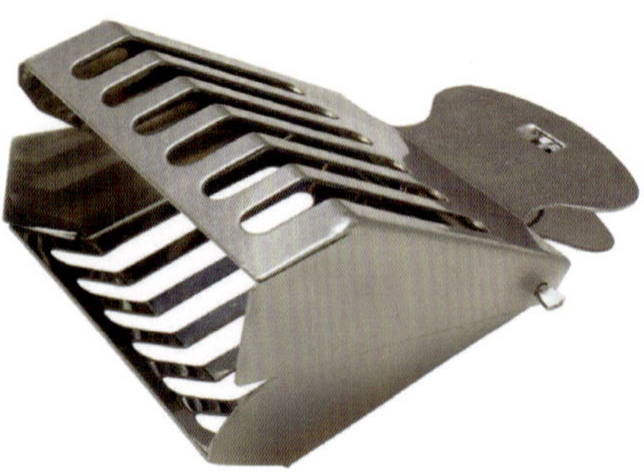

Above: A queen can be held temporarily in a queen catcher. These are usually only used to hold a queen for short periods while the colony is being inspected.

many cages containing queens can be placed into a hive without a free-roaming queen. The bees in this hive will feed all the queens in their cages and control the environment, allowing them to be kept for several weeks/months in this way. An unmated queen can only be kept in a cage for about ten days; after this, her body will have matured, and she will no longer be able to go on mating flights. For shorter periods, including for shipping, it is possible to keep a queen in a queen cage with several workers as helpers and a plug of sugar. Keep these cages in a dark, cool place and provide a drop of water every few days. These cages can be placed in a queen-less hive at the first opportunity so that the bees can eat away at the sugar plug and release the queen. Eating the sugar plug will take a few days, by which time the new queen is likely to be accepted in the new hive. It is generally best to introduce the queen into a small, non-aggressive hive rather than to attempt to place the cage in a large, aggressive hive that may reject the new queen and kill her as soon as she leaves her protective cage. Once the queen lays eggs in the smaller hive, that can be merged with a larger, aggressive hive with much more confidence.

What is the difference between breeding vs rearing queens?

Problem: Rearing queens produces many more queens from a particular stock, while breeding queens implies that the beekeeper intends to undertake a systematic genetic selection programme.

Solution

Professional queen producers often separate the two activities: queen rearing and queen breeding. Some specialise in producing a few high-value selected queens with a known pedigree mated with specific drones (worth several hundred US$ each). In contrast, others will rear hundreds or even thousands of queens for beekeepers nationwide starting from these high-value selected queens but mated under much less stringent conditions.

A beekeeper will want to rear queens from the best possible stock available. As such, beekeepers will start with larvae from a colony with valuable characteristics. This is part of queen rearing. However, if beekeepers actively select the queens over several generations, they are moving into queen breeding. This implies several essential steps:

- **Keeping detailed records of performance parameters such as honey productivity, early spring build-up (essential for pollination of early crops such as almonds), laying behaviour, gentleness and resistance to diseases prevalent in the area.**
- **Controlling or at least influencing the mating behaviour of the queen by, for example, releasing a large number of genetically well-defined drones in the area where the queens will mate, instrumental inseminations or using the services of an isolated mating station where only drones of specific genetics are available to the queens on their mating flights (typically remote islands).**
- **The most challenging part of breeding high-quality, genetically controlled queens is managing and producing drones. The skills and time needed to manage and rear high-quality drones are beyond the scope of a hobby beekeeper.**

CHAPTER 14

Harvesting Honey

Collecting honey supers.

A frame of honey that could win a prize in a competition.

How do I know whether the honey in the hive is ready for harvesting?

Problem: I want to harvest honey from my hive, but I want to know whether the honey is ripe and ready to be harvested.

Solution

Harvesting honey that is not yet ripe (i.e., nectar) is problematic because nectar has a much higher water content (~80 per cent water at the start of the ripening process) compared to honey (<20 per cent water for fully ripe honey). Too much water in honey will lead to fermentation through yeast growing in the diluted honey, converting sugars into alcohol. This is not dangerous but will affect the taste of the honey. Since the amount of water in nectar is very high, there is only a tiny difference between the acceptable water content in honey (20 per cent) and the standard water content (18

Before honey can be extracted, about 80 per cent of a honey frame's cells must be capped. Capped cells are on the left, and uncapped on the right.

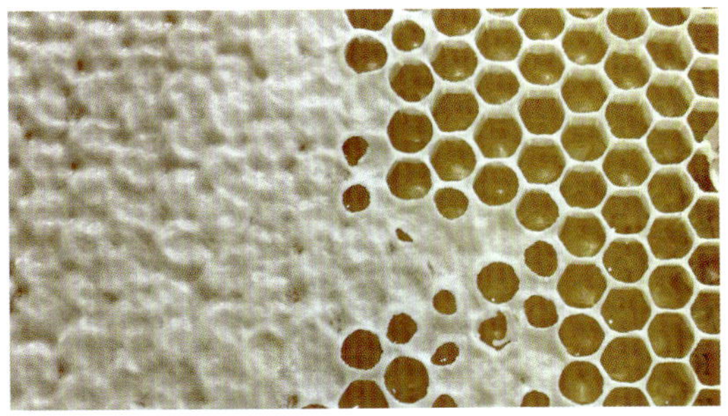

per cent); a small amount of contaminating nectar can quickly reduce the sugar content below acceptable levels.

As a general rule, avoid harvesting frames that have some uncapped honey. If some uncapped honey is present in part of the frame, shake the frame with the uncapped honey facing downwards. If some liquid spills out, the frame should not be harvested. If in doubt, it is better not to harvest a frame with too much nectar and risk having an entire batch of diluted honey that will ferment and be wasted. As a general rule, a frame that has 80 per cent capped honey can be extracted.

How do I harvest honey using a sieve?

Problem: I only have a few frames of honey to extract and want to do it without expensive equipment.

Solution

When starting as a beekeeper, buying extracting equipment is expensive and unnecessary; only a large kitchen sieve is needed. Firstly, cut the capped comb out of the frame in the area under the top frame wire, leaving the comb above it, which joins the top bar of the Langstroth frame. Place the cut-out comb into a large sieve, chop it up with a knife and let the honey slowly drip through the mesh into a container. As the honey needs to be liquid for this type of extraction, it is better to choose a warm day. Alternatively, leave the sieve and bucket in a warm kitchen or in filtered sun to drain slowly. As the honey drains from the comb, periodically chop up the comb and compress it to allow more honey to escape. A plastic bag must be placed over the sieve to stop flies and dust from contaminating the honey. Also, watch for bees that will appear very quickly to rob the harvest if the sieve and honey are not entirely sealed. You mustn't leave the honeycomb in the sun for too long, or the honey will overheat and turn brown, losing much of its flavour and goodness.

When removing sections of comb, cut them into half-frame lengths. Longer lengths are difficult to handle and will likely fall outside the bucket or container holding the sieve, creating a mess.

A simple device to make and use when cutting the comb out of frames is a length of wood 2.5 cm (1 inch) by 5 cm (2 inches) with a screw inserted in the centre of the long side. Choose a screw that is long enough to protrude through the 2.5 cm (1 inch) depth of the wood by at least 1 cm (1/2 inch). The device is placed horizontally across a bucket with the screw point facing upwards. When the side of the frame is balanced on the screw, it can be swivelled easily, and the comb, as it is cut out, will fall into the bucket below. After this, the cut-out comb can be moved from the bucket to the sieve without spilling.

Above: Honey can be extracted using a kitchen sieve. The kitchen needs to be warm to make the honey runny and flow easily through the sieve.

Right: The honey is clean and ready to eat after filtering.

If you leave a strip of comb above the top wire, the bees will use it as a template or starter to build a new comb, and you will not need to attach new foundation to the frame. If you do not add new foundation, the bees take longer to make their foundation and consume additional honey to make the wax. The benefit is a more natural cell size and cleaner comb than foundation purchased with imprinted cell size.

A frame is generally ready to harvest when it is more than 80 per cent full of capped honey. If in doubt, turn the frame on its side and give it a shake — if any runny liquid falls from the frame, it is probably not ready to harvest as the watery honey may ferment, spoiling the rest of the harvest (see 'How do I know whether the honey in the hive is ready for harvesting?').

How do I harvest honey using an extractor?

Problem: I want to extract honey using an extractor.

Solution

An extractor is a centrifuge that spins the frames so that the honey is thrown out of the cells and collected at the bottom of the extractor. Thus, an extractor can only be used when collecting honey from frames and can't be used for collecting honey from top bar hives (see 'What is natural beekeeping?'). Extractors come in various sizes and can be hand-operated or driven by an electric motor. For a hobbyist beekeeper, we recommend a three frame extractor since a two frame extractor is inconvenient, extracting too few frames at once. A three frame extractor is sufficient for up to ten hives. The capped honey frames must first have their wax cappings removed or broken. This can be achieved using a scratching fork, an uncapping roller, or a manual or electric uncapping knife. Once the capping has been scored, pierced or removed, the frames are placed into an extractor to harvest the honey. In the commonly used tangential extractor, frames should be inserted so the bottom bar rotates first, making it easier for honey to flow out.

1. An electric extractor, although expensive, makes extracting a lot easier.

2. A frame of honey ready for extraction.

3. Uncapping a frame using an electric knife.

4. Uncapping a frame using a scratcher.

When determining the number of turns that the frame should be rotated, the 75–150–75 rule applies. First, rotate one side of the frame 75 turns to remove half of the honey from one side. Turn the frame over and rotate the next side for 150 turns to remove all the honey from this side. Now, return the frame to its initial position and rotate 75 turns. This process is used to prevent the frame from 'blowing out.' If, for example, you start by rotating the first side of the frame for 150 turns, the weight of honey on the unspun (inside) side remains the same. In contrast, the side that is being spun (outside) becomes, lighter, more fragile and deforms with the strength of the honey pressing against it from the inside of the frame.

Sometimes, with a thinner foundation, the comb will break during extraction. This is not a problem for the bees; use your fingers to repair the comb so there are no wide splits, return it to the hive and the bees will soon repair the damage, leaving a flawless comb.

If you have only two frames left to extract and use a three frame extractor, keep an already extracted or empty frame in the extractor cage so the unit does not become unbalanced and wobble when rotating. When the two frames are empty of honey, you can remove all three frames, and the extraction is over.

After the frame has been removed from the extractor, it is called a 'sticky' since it is covered in honey and is therefore 'sticky'. If you do not plan to return the stickies permanently to the hive after extraction, they cannot be stored while covered with honey since this will attract bees, ants and other insects. The frames need to be returned to the hive for 24 hours immediately after extracting so that the bees can clean them, removing all the honey remnants, leaving the frames and comb spotless of honey and wax particles. It is good practice to return the frames to the hives they came from to avoid spreading diseases from one hive to another. After 24 hours, you can remove the clean frames permanently from the hive, place them in plastic bags and freeze them for two days. This will kill any eggs of wax moth and small hive beetle. Place the frozen frames in their plastic bags in mouse-proof plastic containers with a sealed lid for long term storage (see 'What should I do when comb is infested with wax moth?').

Should I buy an extractor?

Problem: Extractors are relatively expensive but can be rented or avoided by using alternative harvesting techniques. Here, we discuss the pros and cons of buying an extractor.

Solution

Using an extractor, particularly if you have more than one colony, makes extracting much faster. There are various extractor designs on the market that cater to individual preferences. There are, however, some general points to consider:

- First, is the extractor made from food-grade stainless steel? This type of steel is non-magnetic, so a simple test is to touch the extractor with a magnet and check if the magnet is attracted to the metal. If not, you can be satisfied that the extractor is made from food-grade stainless steel. It is worth noting, however, that some parts of the extractor may not be food grade — for example, the metal gearing. This is not a problem as the gears will not usually contact the honey.
- Does the gate at the bottom of the extractor to release the honey leak when closed? This may not be easy to check but check if the valve shuts tightly. In addition, try opening and closing the gate several times to check if the bolt holding the moveable lid of the gate becomes loose. A further point to check is that the rubber or plastic O-ring sealing the gate entrance protrudes a reasonable distance above the body of the metal or plastic gate, say by 1 mm or more.
- If you plan to extract from shallow frames (see 'What are the different types and sizes of hive boxes?'), ensure they are held in the extractor cage and do not fall through to the bottom of the tank. If the extractor is a good buy in all other respects, this is not difficult to fix. Some stainless-steel frame

wire attached across the base of each frame-holding cage will generally prevent the frame from falling through the cage's base and into the tank. If you spend money on a new extractor and use large numbers of Ideal shallow frames, you may be better off buying an extractor with a cage designed to support both full depth and shallow sizes of frames.

- Check that the frame does not touch the crossbar of the extractor when it sits in the cage, hindering the cage's rotation.
- Consider whether the honey can escape freely from the tank through the honey gate without building up and touching the bottom of the cage. Honey that has built up and touched the bottom of the cage will stop the cage from being turned and put undue strain on the gearing mechanism.
- Can the cage be easily removed for cleaning?
- Ask if the spindle at the centre of the cage sits on a ball bearing at the bottom of the tank. If it does, turning the cage of the extractor will be much smoother and easier. You should also check if replacement ball bearings are available should you lose one while cleaning the extractor.
- What is the warranty period, and are spare parts readily available?
- Are the gears made of iron or nylon? Nylon gears are light but will wear out quickly, and replacements may not be cheap or easy to obtain a few years after you purchase the extractor.
- Many older second-hand extractors are made of galvanised steel. However, honey corrodes galvanized steel, and today, it is discontinued for food hygiene reasons. Take this into account when offered an apparent second-hand bargain.
- Some extractors have plastic drums, which are cheaper than stainless steel drums. We have used plastic extractors, and although we prefer stainless-steel ones, plastic extractors are more affordable and are durable if handled with care.
- The extractor's handle may be mounted on top or on the side. Top-mounted handles may be less expensive than

side-mounted handles, but the difference in cost is insignificant. Try both types of handles and decide which you prefer before choosing one.

- Electric extractors are much easier to use than manual extractors but are significantly more expensive. The motor on the extractor needs to be powerful enough to turn a cage containing full frames without overheating. Electric extractors are a good option if you can justify the cost, as they take all the hard physical work out of extracting. Check that the motor can rotate in both directions at variable speed.
- For safety reasons, the extractor must be earthed, and confirmation must be obtained that it is made to a recognized international electrical safety standard and operates at the voltage suitable for your area (220V vs 110V).

How do I remove bees from my supers when harvesting honey?

Problem: I want to harvest honey from supers, but the supers are full of bees that need to be removed from the frames to harvest the honey.

Solution

Collecting frames with capped honey from the hives for harvesting can be done by removing the bees by shaking and brushing them off. However, this will be time-consuming and result in many bees flying around when harvesting. This will be inconvenient if several boxes are harvested simultaneously in the same apiary. There are several methods to facilitate this procedure.

Escape or clearer board

A clearer board has one or several holes, covered by a funnel-like metal or plastic insert that facilitates the bees going one way but makes it very hard for them to go the opposite way. Hence, when the bees leave the honey super, for example, at night when it is colder and the brood needs to be protected from the cold, they can't return when temperatures increase again in the morning. Hence, after one or two nights, the super is empty of most bees. The remaining bees can easily be brushed off. While this method is easy and effective, it requires opening the hives at least twice on different days, which may not be convenient, especially if the beekeeper must travel some distance to the apiary. When using an escape board, check twice to place it in the correct orientation (we advise writing 'TOP' in large letters on the side facing upwards). Placing the escape board in the wrong orientation on the hive is quite disastrous as it will result in many bees being trapped in the super, exposing the brood and queen to the cold.

Fume board

A fume board relies on a chemical to drive away the bees. The chemical is applied to the fume board, and the board is placed on top of the super containing the honey. The bees do not like the smell of the chemical and are driven out of the box away from the fume board, making it easy to collect the frames of honey with a minimal number of bees that can be brushed off. Check with regulators in your area, as some jurisdictions have banned the use of fumes board with specific chemicals, as there is a small potential for the chemicals to contaminate the honey when it is about to be harvested. Fume boards use chemicals such as Bee-Go, Bee Quick, Honey Bandit Spray or Honey Robber. Bee-Go is based on butyric anhydride, while Bee-Quick is a non-alcoholic blend of natural oils and herbal extracts.

Leaf blower

Some professional beekeepers use leaf blowers (petrol or battery-operated) to remove excess bees from the supers containing honey. This can be done independently or with other methods, such as clearer boards. Typically, the super is placed on its side, and the leaf blower blows air between the frames, forcing the bees to leave the super. The bees are not harmed in the process, and an entire box is emptied of bees in seconds, making this method very effective for large operations where the whole box would be loaded onto a truck for harvesting at a different location.

How long will bottled honey last?

Problem: I harvested honey several years ago and wonder whether it has expired, mainly because it has turned hard (i.e., crystallised).

Solution

When honey is ripe with less than 20 per cent water, its shelf life is at least twenty years. Honey has intrinsic antibacterial properties through the hydrogen peroxide present in it and the high sugar concentration. However, if left open to the air, its hygroscopic properties (i.e., attracting water) will dilute the honey over time. Diluted honey is prone to growing yeast, which will spoil it (effectively turning it alcoholic).

Depending on the source of the nectar and therefore the types of sugar present in the honey, the stored honey may crystallize, becoming hard and opaque. This process will accelerate at certain temperatures. While the texture of the honey has changed, crystallized honey can still be eaten safely.

Can I sell my honey?

Problem: I am producing more honey than I can eat and would like to sell my honey to the public.

Solution

Honey is generally considered a low-risk food product because it tends to remain safe even if stored at room temperature for prolonged periods (see 'How long will bottled honey last?'); it is an ideal product for sale at farmers' markets and other outlets. Nevertheless, some regulations will likely apply to honey made available for purchase. These could include rules around how and where the honey was harvested and bottled, and labelling requirements such as composition information, serving size, daily intake information, batch number, bottling date and seller address. These will differ depending on where you want to sell your produce, so check with local authorities.

Most consumers prefer to buy light-coloured honey, although there are some differences between light and dark honey:

1. Flavour

- **Light Honey has a milder, more delicate flavour, often described as sweet and subtle.**
- **Dark Honey tends to have a stronger, more robust flavour, sometimes with earthy, molasses-like or caramel notes. It may also have a more complex taste depending on the nectar source.**

Honey comes in a variety of colours depending on its floral source. Most consumers prefer lighter-coloured honey. Although there is little difference between light and dark honey, the choice comes down to personal preference.

2. Nutritional Content

- Light Honey usually has less antioxidants than dark honey.
- Dark Honey contains more antioxidants and minerals like iron, calcium and potassium, giving it potential health benefits beyond just being a sweetener.

3. Source of Nectar

- Light Honey is often derived from flowers like clover, acacia or orange blossoms.
- Dark Honey comes from nectar sources such as buckwheat, wildflowers or certain forest flowers.

4. Crystallization

- Honey with more glucose relative to fructose has a tendency to crystallize faster.
- Crystallization of honey can be accelerated by adding a small amount of crystallized honey and mixing the honey

What is the composition of honey?

Problem: I would like to know the composition of honey and how much it varies from batch to batch.

Solution

The composition of honey is consistent, as the bees are doing an excellent job at transforming vastly different sources of nectar into honey that is only marginally different depending on the source of nectar. Over 80 per cent of the honey is composed of sugars. The main sugars in honey are fructose (36–50 per cent) and glucose (28–36 per cent), with sucrose (table sugar) being only 0.8–5 per cent. Variations in sugar content composition also result in honey from different sources crystallizing at different rates. Nevertheless, slight variations occur in mostly minor aromatic components, which give the honey from different sources distinct tastes and colours. Careful analysis of these minor compounds can reveal the origin of the honey and its floral signature. Many people believe that crystallized honey has been adulterated or is otherwise less attractive as food. This is incorrect; only honey that contains higher glucose levels will crystallize, and glucose is better for your health than fructose. Also, people who adulterate honey for profit often add chemicals like sucrose, ensuring the adult erated honey will not crystallize.

How can I determine whether the honey I harvested has a sufficiently low water content?

Problem: **Ripe honey should have a sugar content of around 82 per cent. Honey with a lower sugar content below 80 per cent is likely to ferment, as yeast will grow and convert the sugars in honey into alcohol. Therefore, ensuring the harvested honey has a sufficiently high sugar content is important.**

Solution

Determining the sugar content of honey is easy using a refractometer. This device uses the difference in optical properties of sugars compared to water to accurately measure the percentage of water and sugars in honey. A wide range of refractometers is on the market; in our experience, even the relatively cheap versions work well. However, ensuring the refractometer purchased is made for honey is critical, as the range of sugar/water ratios varies for different applications. Thus, a refractometer made for measuring the sugar content of fruit juice will not work for honey as the scale will not be suitable because fruit juice has a much lower percentage of sugar than honey.

To use a refractometer, place a drop of honey on the slope of the refractometer and close the glass flap, pressing down on the honey to create an even layer. Bring the refractometer to the eye and look through the eyepiece into the light. Use the ring around the eyepiece to sharpen the scale. A line will appear, which should be compared to the scale to measure the percentage of sugar or water.

Extra virgin olive oil calibrates the refractometer and typically measures 71–72 Brix. A Brix of 71.5 corresponds to a water content of 27 per cent. The refractometer has a small screw that allows adjustment of the scale to read the correct percentage of water content (or Brix) when using a reference material. It is essential to calibrate the refractometer before assessing the honey, as the measurement depends on the temperature at which the reading is done, and the refractometer can be misaligned after storage, giving a false reading.

The water content of honey should be between 13 per cent and 18 per cent (i.e., 82 per cent sugar), with a reading above 20 per cent (i.e., less than 80 per cent sugar) being problematic. The most common cause for a low sugar content reading is that the honey harvested is mixed with nectar that is not yet completely ripe.

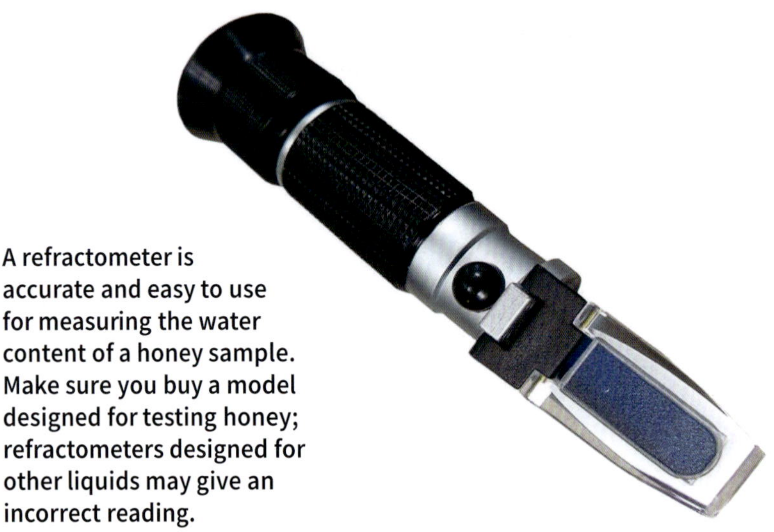

A refractometer is accurate and easy to use for measuring the water content of a honey sample. Make sure you buy a model designed for testing honey; refractometers designed for other liquids may give an incorrect reading.

Can I say that my colony produces honey from a particular flower?

Problem: **I want to produce honey from a specific floral origin and want to know how to approach this.**

Solution

Bees will collect nectar and pollen from a range of floral resources within three kilometres (approximately two miles) of the hive. As a result, in most cases, the honey produced will be a mixture of the nectar collected in the local area. This type of honey is often referred to as 'garden honey'. It is, however, possible to produce honey with specific flavours by encouraging the bees to collect nectar and pollen from particular flower species. This is done by moving the hive to a place where the landscape is dominated by a specific plant at the time of blooming. Thus, during that period, most of the nectar collected will be from a particular flower, with only minor contaminations from other flowers. The result is honey with a consistent flavour and colour, often called floral honey. Some examples include citrus honey, clover honey, buckwheat, lavender, avocado, sage, mānuka honey (a medicinal honey from New Zealand and Australia), leatherwood honey (specific to Tasmania in Australia) and stringy bark (a type of eucalyptus tree).

My colony was infected with AFB or EFB. Can I still eat the honey?

Problem: **My colony has been diagnosed with American or European foulbrood, and I wonder whether I can still eat the honey.**

Solution

Fortunately, there are no pathogens of the honey bee that harm adult humans (honey can contain bacteria called clostridium that can infect infants younger than one year; see 'Can I feed honey to a child less than one year old?'). Bacteria such as AFB, which are lethal to bees, are harmless to people, so eating honey from an infected hive is safe. Management of EFB, and in some countries, AFB, includes the application of antibiotics to the colony. In many countries, minimum withholding periods need to elapse between applying antibiotics and harvesting honey for human consumption to avoid eating honey that contains antibiotic residues.

How do I enter honey in a competition?

Problem: **I want my honey evaluated and compared to other jars of honey in a competitive setting.**

Solution

Many bee clubs and regional, national and international beekeepers' associations organize honey-tasting competitions. In the US, the annual county fair is a treasured institution; almost every county has a fair, and there are usually honey contests and other livestock-related awards. While the rules will change for each of them, there are some general requirements for the presentation of the honey for competition and criteria against which the honey will be judged. Typically, the honey will be bottled in a plain glass jar of a specific volume/weight without a label. There are generally different categories for each honey (garden honey, different floral kinds of honey relevant to the region), and a variety of others, such as honey produced by an amateur, a beginner beekeeper, creamed honey, etc.... The judging criteria are published and often involve taste, clarity, viscosity and lack of bubbles or foam on the surface.

The judges are often professional honey sommeliers who use a wide array of descriptive terminologies to describe the properties of each honey. While monetary prizes are to be won, many beekeepers participate to obtain the 'bragging rights' associated with winning a fiercely contested local competition.

Where can I extract my honey?

Problem: **I want to safely extract honey where bees can't access open frames of honey.**

Solution

To minimize the number of bees attracted to your honey, extract them in an enclosed space such as a garage or kitchen or after dark. When extracting during daylight, there must not be an open entrance to the extracting room, or you will soon be swarmed with bees you cannot remove. If you cannot extract in a closed area, extracting after dark when bees do not fly is a good alternative. Use red light that bees can't see to minimize attracting bees if harvesting near the apiary.

Can I feed honey to a child less than one year old?

Problem: **I have heard that feeding honey to a child under one year old is not recommended, and I wonder why.**

Solution

Infant botulism is caused by *Clostridium botulinum* spores, sometimes found in pasteurized and unpasteurized honey. When an infant ingests honey, bacteria from these spores can grow and produce toxins that may lead to paralysis. The immune systems of older children and adults are usually sufficiently strong to fight infections by this pathogen, so there is no problem in these cases.

CHAPTER 15

Natural Beekeeping

A bee collecting pollen and nectar.

What is natural beekeeping?

Problem: **As a new beekeeper, I would like to manage my bees naturally. I wonder what natural beekeeping is and how to keep my bees naturally.**

Solution

Natural beekeeping is an approach to keeping bees that emphasizes working with bees and nature rather than trying to control or manipulate them. It typically focuses on practices that support the health and well-being of the bees, minimize human intervention, and maintain a balance with the local ecosystem. Some fundamental principles of natural beekeeping are:

Minimal intervention: Natural beekeepers disturb the hive as little as possible, avoiding routine inspections, which can stress the bees. Instead, they rely on observing colony behaviour at the hive's entrance and only intervening when necessary. However, in most jurisdictions, there are minimum biosecurity requirements for inspecting the brood for signs of diseases.

Inspecting a top bar hive.

Respect for the bees' needs: This involves providing them with a habitat that supports their natural behaviours. However, in many countries, it is a legal requirement to be able to remove comb from the hive to inspect the brood for diseases.

Chemical-free management: Natural beekeeping avoids synthetic chemicals and antibiotics, instead focusing on integrated pest management (IPM) to control pathogens. These can include the following:

- **Physical controls**: Using mesh screens or traps prevents pests like *Varroa* mites from harming the bees.
- **Biological controls**: Introducing natural predators or competitors to control pest populations.
- **Cultural practices**: Ensuring proper hive hygiene, using good hive designs and managing colonies to reduce the likelihood of any possible pest outbreaks.

Swarm Management: Swarming is a natural part of a bee colony's life cycle. Natural beekeepers often allow swarms to occur rather than attempting to prevent them. Depending on the local area there may be restrictions in allowing the bees to swarm and legal ramifications if the beekeeper does not prevent swarming.

Sustainable practices: This approach emphasizes sustainability, ensuring beekeeping practices do not harm the environment or over-harvest honey. Natural beekeepers often leave colonies with sufficient honey to last over the winter, only harvesting what is left, above what the bees need for their consumption. Commercial and hobby beekeepers frequently take most of the honey and feed syrup as food to overwinter.

Local adaptation: Natural beekeepers may focus on breeding bees better adapted to local conditions and more resistant to local diseases and pests than importing queen bees from other regions.

Hive design: Natural beekeepers often use hive designs that cater to the bees' natural behaviours. For instance:

- **Top bar hives**: These hives use horizontal bars instead of vertical frames. They mimic natural hive structures more closely, allowing bees to build their comb and manage their hive. The bars can be removed to

1. A Warre hive.

2. A top bar hive is a favourite with natural beekeepers.

3. Wood shavings, called quilts, are used to insulate a Warre hive.

inspect the comb for diseases but require more careful handling than traditional wired frames.

- **Warre hives**: These vertical hives are designed to mimic the natural tree cavities that bees might choose in the wild. They emphasize minimal intervention and allow bees to build their comb in a more natural arrangement. Warre hives are typically expanded by adding boxes at the bottom of the stack, so the bees move the brood down and add honey above, which more closely mimics the natural situation.
- **Sun hives**: These are complex round hives hanging from a wooden platform and woven straw that can be opened by removing the bottom half to access the comb with minimal disturbance to the bees.

Colony health: Maintaining colony health is a cornerstone of natural beekeeping. This involves:

- **Ensuring adequate nutrition**: Providing supplementary feeding only when essential and ensuring the bees access a diverse range of nectar and pollen sources.

- **Promoting natural behaviours:** Allowing bees to engage in natural behaviours, such as building comb in their preferred patterns and swarming, can reduce stress and improve colony resilience. In many countries, however, there are legal requirements to prevent swarming.

Holistic approach: Natural beekeeping often embraces a holistic view of the ecosystem, understanding that bees are part of a more extensive web of life. Practices might include planting bee-friendly plants, creating habitats for other beneficial insects and supporting biodiversity in the surrounding environment.

Bee behaviour observation: Natural beekeepers strongly emphasize observing and learning from the bees' behaviour rather than relying on pre-set schedules or routines. This observation helps them more intuitively understand the colony's needs and health.

Natural beekeeping aims to create a more harmonious relationship between humans and bees. It fosters an environment where bees can thrive while allowing beekeepers to sustainably enjoy the benefits of beekeeping, such as honey and wax.

What are the advantages and disadvantages of a top bar hive?

Problem: I am deciding whether to buy a top bar or Langstroth hive to start beekeeping. I would like to know the advantages and disadvantages of using a top bar compared to a Langstroth hive.

Solution

Top bar bee hives, often called horizontal top bar hives, are an alternative to the traditional Langstroth. There are two versions of the hive: the Kenyan and Tanzanian hives. The Kenyan top bar hive has sides that slope towards the bottom at about 60 degrees, while the Tanzanian top bar hive has vertical sides. They offer a unique approach to beekeeping; they have advantages and disadvantages like any system:

Advantages:

- They are generally less expensive to build than Langstroth hives, although they may be more expensive to buy as an assembled hive. The design is simple and requires fewer materials than a Langstroth hive. A hive is easy to make from scrap wood or wood inexpensively purchased from a builders' supply store.
- No heavy lifting is required when inspecting or managing the hive; only the lid must be removed, making inspections easier to manage.
- The horizontal top bar hive can be built so that the top is at waist level, simplifying inspection of the hive. In some cases, this would allow wheelchair users to practise beekeeping.
- Access to brood or honeycomb is easy. This is because both the brood chamber and the honey chamber (super) are at the same level, and the beekeeper does not need to remove the super to gain access to the brood chamber.
- A single horizontal top bar hive can house two separate colonies using an internal partition, usually called a follower or divider board, suitably positioned to keep them apart, with an entrance at the bottom for the lower brood box, and an entrance at the top for the upper brood box. Many variations of top bar hives allow beekeepers to choose a design that fits their needs and preferences.
- Top bar hives encourage bees to build their comb naturally, without using frames. Natural beekeepers believe this leads to healthier and more natural comb structures, benefitting the bees' overall well-being.

Disadvantages:

- Unlike the Langstroth hive, these hives do not use standardized parts, and a beekeeper may have difficulty purchasing a ready-made or kit, such as a horizontal top bar hive. If needed, beekeepers using top bar hives may have to make parts, such as a divider board or queen excluder.
- The hive needs to be placed on a level surface, and due to its large footprint, this may be more difficult than levelling a Langstroth hive.
- Due to its larger size, moving a horizontal top bar hive may be more complex than moving a Langstroth hive.
- Managing pests and diseases can be more challenging due to the hive's design and the difficulty of removing and inspecting combs.
- Comb built in a top bar hive is more fragile than comb built in a frame-based hive.
- The hive's design can make managing swarming behaviour more challenging. Proper swarm management, a legal requirement in many areas, requires careful monitoring and intervention.
- Harvesting honey can be more challenging than for Langstroth hives. Since the combs do not use built-in frames or wire support to run through the comb, the comb must be cut out to extract honey. Comb may be attached to the side of the top bar hive, making it difficult to remove without destroying the comb while it is still inside, making a mess.
- Top bar hives typically have less internal space than Langstroth hives, which can be increased by adding additional brood boxes and supers. This may limit the colony's honey and require more frequent management.
- The overall size of a top bar hive cannot be changed. However, the space available to bees can be restricted using a follower board.

Choosing a top bar hive often depends on your needs, beekeeping goals and preferences. They can be an excellent option for those who value a more natural approach and are willing to handle the unique challenges they present.

How do I inspect a top bar hive?

Problem: **I recently started beekeeping using a top-bar hive and want to inspect the colony for health, honey and disease.**

Solution

Inspecting a top bar bee hive differs from inspecting traditional Langstroth hives, but the process is straightforward once you get the hang of it. The most significant difference is that top bars may be difficult to remove without damage to the comb since they do not have sides or wire support. The comb may also be securely attached to the side of the hive, making removing a top bar difficult without damaging the comb.

- Remove the roof of the top bar hive. In some cases, the roof is hinged to allow easy opening.
- The queen usually lays eggs on top bars near an entrance, and honey is stored further away. Gently tap along the side of the hive, moving away from the entrance, until a hollow sound is heard. Gently lift the top bar after the start of the hollow sound. Removing the first and often the second end bars may be difficult if some comb is attached to the hive wall. A long knife will need to slide down the side of the hive to cut the comb away from the wall. In our experience, we have never seen a top bar hive in which the very end top bars have contained foundation, making them easy to remove. With a Langstroth hive that uses frames with four sides, the first frame may be removed and placed on the ground,

Inspecting a top bar hive may be more difficult than inspecting a Langstroth since comb is not supported, may break off and may be attached to the side of the hive.

resting against the Langstroth hive. Since a top bar does not use sidebars or a bottom bar, if the first frame needs to be stored away from the hive, allowing easier access to the remaining top bars, a separate holder or support is required. A system of hooks on the outside of the hive is sometimes used to store the top bar removed from the hive temporarily.

- Carefully lift each bar out of the hive one by one. Be gentle to avoid breaking the comb or disturbing the bees too much.
- Check each comb for the queen's presence. If you cannot locate the queen, eggs or larvae, move on to the next step.
- Look for tiny eggs laid in the bottom of the cells. Larvae will be curved and creamy, and pupae will be in sealed cells. The presence of eggs or young larvae indicates the presence of a queen.

- **Ensure enough honey and pollen are stored. The honeycomb should be capped, and pollen should be stored near the brood.**
- **Look for pests such as *Varroa* mites or signs of disease. Common symptoms include deformed wings, abnormal brood patterns or mouldy combs.**
- **Observe the bees' behaviour. If they seem unusually aggressive or passive, there might be an issue.**
- **Once you've completed your inspection, gently place each top bar back in its original position. Avoid damaging the comb or displacing the bees.**
- **Put the roof back on the hive. Ensure it fits well to protect the bees from the elements.**

It's a good idea to keep a log of what you observe during each inspection, including the condition of the hive, the presence of the queen and any issues you notice.

Inspecting a top bar hive requires a bit of finesse, but it becomes easier and intuitive with practice. Practice will help you keep your hive healthy and productive.

How do I harvest honey from a top bar hive?

Problem: My top bar hive is full of honey, and I need to harvest honey from it.

Solution

The main difficulty with extracting honey from a top bar hive is removing the top bars to collect comb. Care must be taken since the honeycomb is heavy but has little strength without the wooden sides of the supporting wire threaded through it found in Langstroth frames. One less concern is that since you permanently remove the comb to extract honey, the comb can be broken and taken away while the top bars are removed.

Since the top bars with the comb are delicate, removing the honeycomb must be done near the hive. Using a plastic container with a bee-proof lid, remove each top bar and cut the comb into the container, leaving about 1 cm of comb attached to the top bar for workers to use as a template when making new comb. Quickly close the container after adding the comb from each top bar before returning the top bar to the hive, where workers clean honey and start rebuilding comb. Push any top bars containing partially capped honey you didn't harvest towards the front or brood area and put the top bars that honey was harvested from behind those; that way, they will finish the already-started ones before starting over on the old ones.

Any damaged comb left in the hive can be removed and placed in the container. If the comb on a top bar to be removed is attached to the side of the hive, use a long knife, such as an uncapping knife, to slide down the side of the hive and cut the comb from the hive's walls. Return all top bars, close the hive and you are ready to extract honey.

Since the comb in the container is broken, an extractor cannot be used. Place all the comb into a nylon, net or wide mesh bag and let the honey drip into a container. Alternatively, place the comb into a flour sieve and let the honey drip into a container. Every half hour or so, cut up the comb so more honey can be freed from the comb. Place a plastic bag over the sieve and container into which the honey flows to stop bees from feeding off the extracted comb. The leftover comb still contains a lot of honey. So that it is not wasted, it can be fed back to the colony by placing it under the lid or on the floor of the hive so that workers can access it. See 'How do I harvest honey using a sieve?'

A more efficient way to extract honey from comb is to use a fruit press, typically used to crush apples to produce apple juice. A fruit press is expensive, but it makes extracting easier and harvests more honey. Correctly used, the remaining wax from a fruit press is compressed in a block with little honey left.

How do I combine two top bar hives?

Problem: **I have two top bar hives, one of which is weak and must be combined.**

Solution

These methods are similar to how Langstroth hives are merged in the section: 'I have two weak colonies; how can I merge them?'

Two methods exist for combining colonies from separate top bar hives or merging a swarm with a colony in a top bar hive. Apart from the design of the hives, they are the same as used with Langstroth hives:

Direct merge

- Both hives must be near each other to ease moving top bars.
- Decide which colony is more robust and remove the queen from the weaker colony.
- Do not kill the weaker queen now, as the stronger queen may accidentally be killed during the merger.
- In the strong hive, ensure sufficient space at the end to add top bars from the weaker hive.
- Gently move each top bar from the weaker hive to the stronger hive, keeping the order of top bars the same.
- When merging two colonies, there needs to be a second entrance to the top bar hive at the opposite end so the added colony can access its part of the hive.
- To increase the probability that the two colonies will merge without fighting, dust workers from both colonies with finely powdered icing sugar or spray them with sugar-water. Workers will be more concerned with collecting sugar than with fighting.
- Slowly replace the lid.
- The disadvantage of the direct merge method is that the two colonies may fight, severely depleting the colony size.

- After one week, check that the two colonies have merged; this should be evident. If there are many dead bees outside the entrance, the two colonies have not mixed well, resulting in fighting.
- If the strong queen is still alive, destroy the weaker queen.

Newspaper method

- The newspaper method is slightly more complicated but more likely to result in two colonies merging successfully.
- Both hives must be near each other to ease moving a brood box or frames.
- Decide which colony is more robust and remove the queen from the weaker colony.
- Do not kill the weaker queen now, as the stronger queen may be killed during the merger.
- Decide which hive will remain in the exact location.
- In the strong hive, ensure sufficient space at the end to add top bars from the weaker hive.
- Place a plastic queen excluder designed for the top bar hive between the existing colony and where the second colony will be located. The plastic queen excluder may need to be cut to size.
- Attach a sheet of newspaper to the queen excluder so the workers from both colonies cannot fight each other. The two colonies will take about one or two days to remove the paper and merge.
- There needs to be a second entrance to the top bar hive at the opposite end so the added colony can access its part of the hive.
- Some beekeepers slit the newspaper so that the bees can easily chew the paper, allowing the workers to merge.
- Place top bars from the second brood box in the hive and replace the lid.

- **The bees should chew their way through the newspaper.**
- **After one week, check that the two colonies have merged by removing the lid and seeing if the newspaper has largely disappeared.**
- **If the strong queen is still alive, destroy the weaker queen.**

Swarms

If a swarm that has not yet been housed in a top bar hive is to be merged with an existing colony, either of these methods can be used. Ensure the back of the hive includes sufficient top bars to settle the swarm and allow them to build comb. The top bars may either be new or contain some comb, including brood and honey, to make the hive more attractive to the swarm.

What are long hives?

Problem: I like the horizontal hive, which allows beekeeping without lifting heavy boxes and inspecting the hive from a sitting position. Still, I would like to use standard Langstroth frames because of their ease of manipulation and honey harvesting using an extractor.

Solution

Beekeepers who want to combine some of the advantages offered by standard Langstroth frames with the ease of manipulation offered by horizontal hives can make a horizontal hive that accommodates standard frames. In building these hives, great care needs to be taken to respect the bee space central to the Langstroth concept. Some plans can be found on the internet, but these hives may be difficult to purchase commercially. Horizontal hives that contain top bars instead of frames are called Tanzanian Top Bar hives.

A long hive is more straightforward to inspect since there are no supers to remove, and access to all parts of the hive is at waist level.

Schematic of a long hive

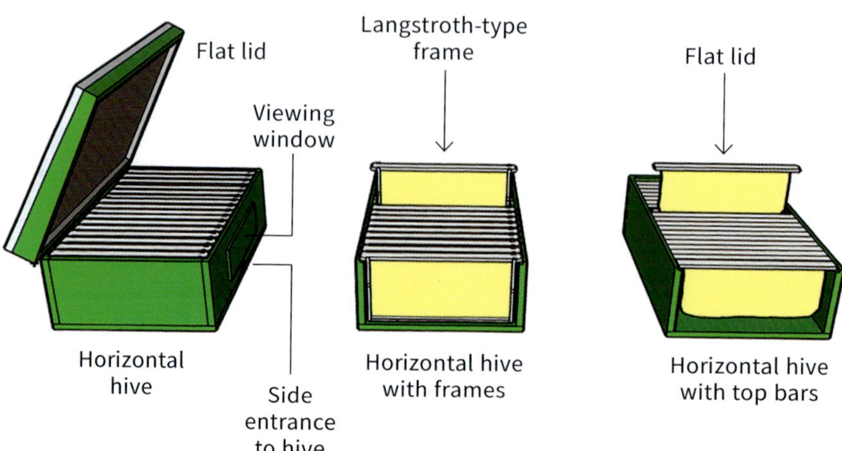

How do I split a top bar hive?

Problem: I have a strong top bar hive that may soon swarm and want to split the colony.

Solution

Splitting a top bar bee hive is used to manage bee populations and create new colonies, including managing colonies that may soon swarm. The steps to be taken are as follows:

1. Inspect the hive:

- Smoke the entrance to calm the bees.
- Open the hive and gently inspect the combs. Look for queen cells and brood patterns. Ideally, you want to split when the hive is preparing to swarm or has a strong queen.

2. Split the hive:

- Set up a new hive body or box a reasonable distance from the original hive (at least ten metres (yards), but further is better).
- Identify combs with brood, honey and pollen in the original hive. Moving some of these combs to the new hive would likely result in a successful split.
- Carefully move selected combs from the original hive to the new hive. Make sure each hive has a good balance of resources and comb.
- If possible, find a queen cell or a frame with eggs.
- If you've found the queen, put her in the new hive or leave her in the original hive. If you leave her, ensure the new hive has queen cells or eggs to raise a new queen. See Chapter 13, 'Queen Rearing'.
- If there are insufficient top bars in either hive, add as many as possible. Either new top bars or top bars with comb may be added.

- **Close both hives.**
- **Monitor both hives for a few weeks. Check for egg-laying and overall activity to ensure the colonies are functioning well.**
- **Ensure they have enough food and space to grow.**
- **Splitting a hive is a great way to manage bee populations and prevent swarming, and with practice, you'll get more comfortable with the process.**

See 'How do I split a hive?' Splitting a top bar hive is similar to splitting a Langstroth hive.

CHAPTER 16

The FLOW Hive

Flow hives in a garden.

What is a Flow Hive and how does it compare to a Langstroth hive?

Problem: I would like to know the main differences between a Flow Hive and a Langstroth hive and what I should consider when deciding whether a Flow Hive suits my needs.

Solution

The Flow Hive was developed in Australia by a father and son team and is a relatively new development in the way beekeepers can harvest honey from the hive. Flow Hives are compatible with eight- or ten-frame Langstroth hives because the bees deposit the honey in a super (i.e., a hive box storing honey above the queen excluder) containing special plastic frames. These frames are designed to be wider than normal Langstroth frames (i.e. an eight frame box contains six Flow Frames,

Flow Hive in a garden.

while a ten frame box has seven Flow Frames).

The honey deposited in Flow Frames can be harvested without opening the super and removing the frames. Indeed, Flow's split cell technology allows the cells to be opened, and the honey flows out of a tube at the bottom of the frame directly into a jar.

The Flow Hive is widely used and works well especially if the honey harvested is liquid, which is generally the case. Still, under specific circumstances where honey crystallises in the hive, Flow Frames do not work well, and special precautions must be taken to allow harvesting. Therefore, before considering operating a Flow Hive, assess whether your Flow Hive will be used in an area where the bees produce honey that crystallizes easily (e.g. honey produced from canola flowers). It should be noted that crystallized honey in frames also presents a problem when harvesting using more traditional methods such as centrifugation.

Another important consideration is cost. Flow Hives and especially Flow Frames are substantially more expensive than normal Langstroth frames. However, purchasing a centrifuge for extraction is unnecessary when using a Flow Hive. Thus, for a relatively small number of hives (i.e, up to two to three), the savings from not purchasing a centrifuge compensate for the higher costs of Flow Hives. Thus, an important consideration is the total number of hives that a beekeeper intends to keep. The more hives a beekeeper has, the cheaper it is to use a centrifuge to harvest honey.

One benefit of a Flow Hive is that harvesting is easier for the beekeeper and gentler for the bees. Beekeepers no longer have to open their hives and remove all the frames with capped honey to harvest, which usually results in many bee fatalities and is stressful and disturbing for the colony. A Flow Hive is carefully designed to ensure the bees remain safe and undisturbed while the honey flows out of the hive and into your jar.

Another factor is the convenience of harvesting, especially if regularly harvesting small amounts of honey is of interest to you — an advantage of a Flow Hive is the ability to quickly harvest small amounts of honey from a few hives. Indeed, it is possible to harvest one or even a fraction of one frame at a time and repeat this process every few weeks during a strong honey production season. In contrast, harvesting with a centrifuge requires preparation and a substantial clean-up operation. It is generally undertaken only once or a few times yearly (unless you have more than 50 hives). This will be important for some people as they can collect a single pot of honey for direct consumption or as a gift to a visitor who can also watch the honey flow out of the hive into the jar.

It is important to note that beside harvesting, the general procedure for keeping bees in a Flow Hive remains very similar to that of the Langstroth. Indeed, both hive types require similar procedures for swarm control, inspection and treatment for diseases, requeening, splitting and any other tasks outlined in this book.

Which Flow Hive should I buy?

Problem: I have decided to buy a Flow Hive and have to choose from several available designs.

Solution

Starting as a beekeeper with a Flow Hive is the same as starting with a Langstroth hive. A brood box and bees are needed. As the colony grows and needs to expand, a Flow Super containing Flow Frames is placed above the brood box. Worker bees use Flow Frames to store honey, which, when ripe, can be extracted with minimal disturbance to the colony.

Flow Frames are designed to extract at the hive without needing to open the super and take out the frames. This is very different from traditional Langstroth hives, where the super is opened, honey frames are removed, taken to the garage for extraction and then returned to the hive, which requires reopening and further disturbance of the colony. Using the Flow Frames eliminates this time-consuming, disruptive and messy honey extraction process. Since there is little disturbance to the bees, the likelihood of being stung during harvesting may be significantly reduced.

Design of the super

Although you can buy a complete Flow Hive, the unique part is the super with Flow Frames. The remainder: base, brood box and roof (or lid), are variations of traditional Langstroth hives, although Flow's creativity makes the final product more attractive and easier to manage than a conventional hive.

The super that holds the Flow Frames has a different back end than a Langstroth super. Because Flow is designed to extract honey directly from the super without removing any frames, the back provides access to frames via a removable panel, which is kept in place by a latch. Removing the panel allows you to attach tubes to the bottom of the frames through which honey can be extracted. At the top of the rear of the super is another removable strip of wood through which you insert a metal Flow Key into each frame. There are two side windows on the super, and together with the rearview, they can be used to check how much honey is present and if it needs extracting. Flow Frames are designed to extract honey without removing them from the super. When the key is inserted into the top and turned, the cells in the frame move apart, allowing the honey to flow to the bottom. The honey flows into a trough at the bottom and out of a tube at the back into a container. Although a Flow Hive comes with one Flow Key or lever, buy a second Flow key. Insert both keys into the hole, turn one clockwise and the other anticlockwise. This makes opening the Flow Frame for extracting easier.

Which design of the Flow Hive should I buy?

You can buy a stand-alone Flow Super with six or seven Flow Frames, replacing traditional supers with standard eight or ten Langstroth frames. You can also buy a hybrid super with three Flow and four standard Langstroth frames to collect fresh honeycomb. The hybrid system also allows the rotation of brood frames into the super so that fresh brood frames can be added more easily (see 'What should I do when the honeycomb turns hard and dark?'). A Flow Super has the same internal dimensions as a traditional eight- or ten-frame Langstroth hive, allowing it to be used with conventional brood boxes.

The big choice occurs when you buy a complete hive, including a base, brood box, super and lid. There are four designs: Flow Hive 2+, Flow Hive 2, Flow Hive Classic and Flow Hive Hybrid.

	Flow Hive 2+	Flow Hive 2	Flow Hive Classic	Flow Hive Hybrid
Timber	Western Red Cedar	Araucaria	Araucaria	Araucaria
Flow Frames	6 or 7	6 or 7	6 or 7	3 Flow Frames, 4 conventional frames
Painting not included	Can be oil stained or painted	Outdoor paint	Outdoor paint	Outdoor paint
Removable multifunctional tray at the back	Included	Included		
Observation windows	Two	Two	One	One
Entrance Reducer	Included	Optional	Optional	Optional
Ant Guard	Included	Optional		
Adjustable hive stand	Optional	Optional		
Base with ventilation control. Designed for pest management.	Included	Included	Sloped Base-board	Sloped Base-board
Multifunctional tray	Included	Included		
Including spirit levels for hive set-up	Included	Included		
Brood Box with eight or ten frames	Included	Included	Included	Included
Inner cover	Included	Included	Included	Included
Gabled roof	Included	Included	Included	Included

TABLE 5: Flow Hive shows that Flow Hive comes in various designs, with each improvement designed to make beekeeping more accessible and more convenient. Of course, if you are an existing beekeeper and already have a conventional Langstroth hive, you can save money by only buying a Flow Super and possibly upgrading later.

How do I keep bees with a Flow Hive?

Problem: **I just bought a Flow Hive and want to know how managing the colony differs from a conventional Langstroth hive.**

Solution

Starting as a beekeeper with a Flow Hive is the same as starting with a Langstroth hive. A brood box and bees are needed. This part is the same as described in Chapter 1, 'Getting Started'. As the colony gets stronger and needs to expand, a Flow Super that contains Flow Frames is placed above the brood box, preferably above a queen excluder. If you want to speed up the process, you can encourage the bees to use the plastic Flow Frames with a thin layer of molten wax applied to the surface of the frames with a roller. Worker bees use Flow Frames to store honey instead of conventional frames with wax or plastic foundation, which, when ripe, can be extracted with minimal disturbance to the colony.

If you have a lot of honey coming in or live distant from your hive and cannot check it regularly, you can use two Supers: one Flow Super and one containing Langstroth frames on the same hive. The authors would put the Flow Super above the brood box to encourage workers to fill Flow Frames first. Towards the end of autumn, as the number of bees in the colony decreases and nectar is in short supply, you may need to reduce the size of your Flow Hive from two boxes to one box ready for winter, called packing down. Any frames full of honey can be extracted, washed in water and stored. Another way to clean a frame is to leave it in warm, soapy water. The honey will dissolve over time, and the frame can be rinsed with cold water.

How do I extract honey from a Flow Hive?

Problem: I just bought a Flow Hive and want to use it to harvest honey from the super.

Solution

Although simpler than conventional extraction methods, a Flow Hive still requires care and attention. Frames should be extracted when they are about 80 per cent full of capped honey. Since extraction is performed without opening the hive, you need some experience determining which frames are full and ripe. In the first year, beginners should regularly open the hive to inspect the brood and the super. The brood is inspected for colony health, and the Flow Frames are examined to determine the ripeness and amount of honey present. After a season, a Flow Hive owner can gauge the ripeness and amount of honey by using the hive's weight and the views through the rear and side windows.

If an extracted frame is not 80 per cent capped, there is a risk the honey will be too watery and will ferment. Open the side viewing panel, and if the end frame is 80 per cent capped, probably all frames are capped since end frames are usually the last ones to be filled. If the end frame is not capped, you need to open the hive, remove a central frame, and check how much it is capped. If only one frame is capped, you can extract that. Otherwise, you might need to wait a week and check again.

Follow these steps when extracting:

- **Before extracting, at the back of the hive, remove the key access cover at the top of the super and the rear door. Remove the honey gate trough cap at the bottom of the frame and insert the clear plastic tube into the hole at the bottom. Place a large jar to collect honey underneath the end of the tube.**
- **Next, insert the long Flow Key into the top hole, inserting it under the bar across the centre of the hole into the bottom slot. To open the frame, insert the key below the bar; to close**

the frame, insert the key above the bar. Turn the key slowly, clockwise, or anticlockwise; it does not matter which. Once the frame has been opened, honey will start running into the channel and, from there, into the jar. It is possible to harvest only the front of the frame by inserting the key partway into the slot, leaving the area at the back of the frame full of honey. This can be useful if you only want to harvest a small amount of honey and leave more for the bees.

- After the honey has stopped running, insert the key above the bar across the top and turn it 90 degrees. This will realign the cells to their hexagonal shape. Leave the key in this position for a minute or two to ensure the cells fully return to the closed position. Then, replace the honey trough cap, which is designed not to be inserted if the frame is not closed.
- Remove the plastic tube and replace the stopper. There will still be a small amount of honey in the channel at the bottom of the frame. Frames have a small 'Leak Back Gap' in the channel, so residue honey will escape into the brood box, where bees gather and store it. Worker bees will now clean and repair the frame and return to storing honey.
- Repeat this for each frame that needs extracting, then return the top strip and back cover to their original positions. If there is more than one frame to extract, you could extract more than one frame at once. One Flow Frame gives about three kilograms of honey, about the same as a Langstroth frame.

It is good practice to measure the water content of extracted honey. The water content can be measured using a refractometer (make sure you get one for honey, and not for fruit juice or other applications) and should be no more than 20 per cent water (see 'How can I determine whether the honey I have harvested has a sufficiently high sugar content?').

FLOW® FRAME

The Flow® Frames fit into a standard Langstroth super (eight or ten frame).

Two simple cutouts in one end of the super allow access for honey collection and end frame observation.

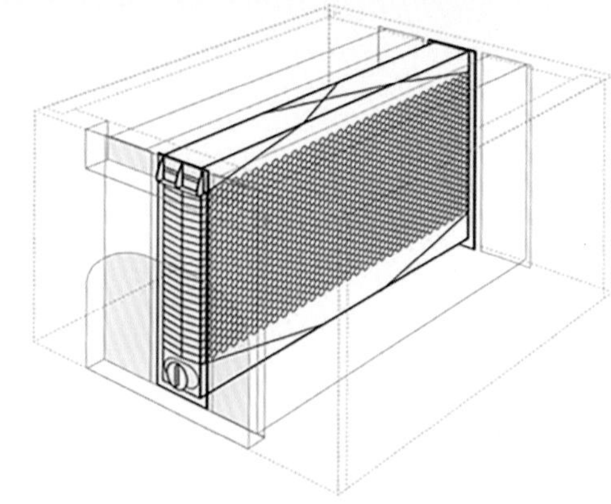

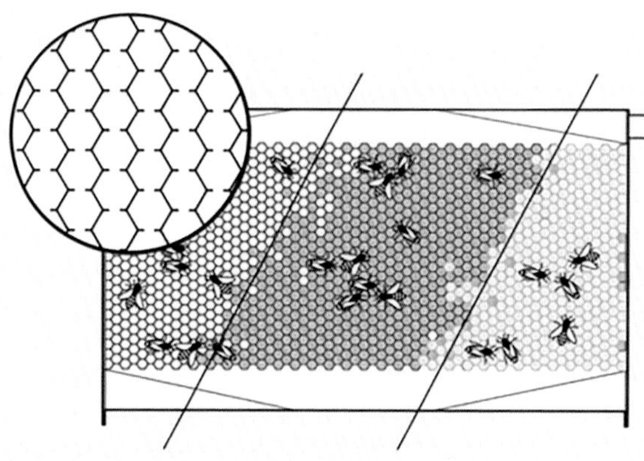

The Flow® frame consists of partly formed honeycomb cells.

The bes complete the comb with their wax then fill the cells with honey, before finally capping the cells.

When the frame is full, it's ready to harvest.

1 Remove the key access cap
2 Insert honey tube into honey trough
3 Insert Flow key into bottom slot
4 Rotate Flow key 90° downwards

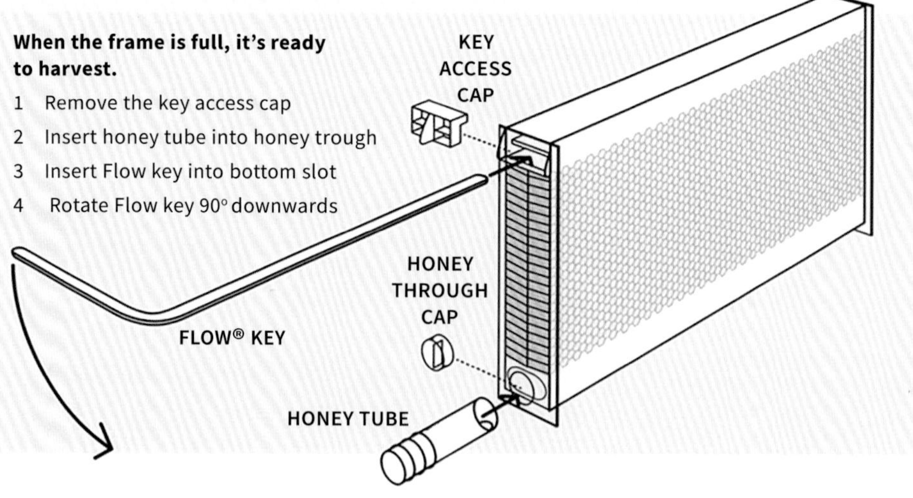

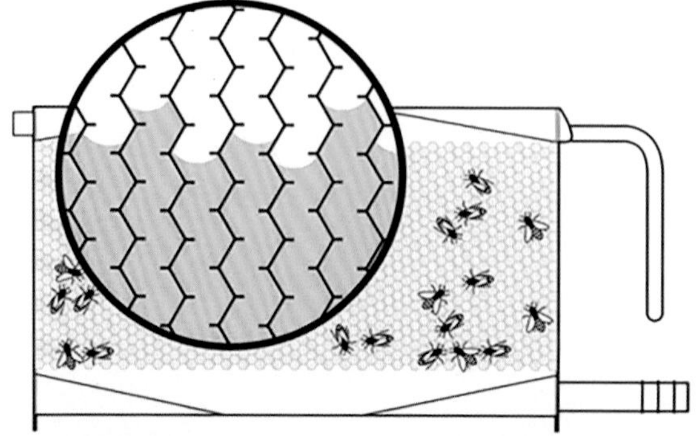

Inside the honeycomb the cells have now split and turned into channels for the honey to flow down.

The bees remain undisturbed on the surface of the comb. If there does happen to be a bee down an empty cell, it won't get injured as there is enough space between the comb walls.

1. Use two keys to extract honey by opening (splitting) a Flow frame. The use of two keys makes splitting the frame easier. Buy the second key when you order the Flow Hive, this will save on postage later.

2. Extracting honey from a Flow frame straight into a jar.

3. Several jars of honey being filled at the same time.

Opposite: The design of a Flow frame ensures that honey can be extracted easily without the mess often associated with extracting from a Langstroth hive or top bar hive.

CHAPTER 17

Other Products from Your Hive

Uncapping a honey frame with a hot-knife.

Can I eat comb honey?

Problem: I want to eat the honeycomb and wonder what precautions I should take in producing honeycomb.

Solution

Many decades ago, honey was generally sold as comb. This made mixing in some sugar syrup much harder, guaranteeing that the product sold was indeed honey. However, honeycomb is no longer sold in many parts of the world except in specialty stores or farmers markets, with the public preferring to buy honey in jars.

Consuming honeycomb is still perfectly ok, but some precautions are essential. The wax component of the honeycomb can be eaten, but it is best if the bees entirely make the comb and does not contain foundation, especially if the origin of the foundation is unclear and one can't guarantee that it is not contaminated with chemicals used in the hives (miticides, antibiotics,

Pollen comes in many colours.

Many people enjoy eating comb honey, which is regarded as a delicacy in many parts of the world.

etc.,) or from the environment (pesticides brought in by the bees). Thus, if you plan to eat or sell honeycomb, make sure that you only use a starter strip of foundation. Other precautions include using a queen excluder to prevent the queen from laying in the comb that will be eaten. Indeed, if the comb has been used for brood, there will be residues of brood in the comb, making it dark and unpalatable due to the presence of cocoons and other leftovers from the brood-raising process.

How can I produce honeycomb to be eaten?

Problem: I want to produce some honeycomb for my consumption or sale.

Solution

The production of honeycomb requires a very strong nectar flow as the bees must produce new comb, fill it up with nectar, and transform it into capped honey over a short period of time. Indeed, to obtain a capped honeycomb with an attractive white capping, the honeycomb can't have been in the hive for too long, or it tends to become brown due to bees continuously walking over it. In addition, the entire frame needs to be capped so that it is easy to cut out into square sections without worrying about having some uncapped areas.

The presentation of honeycomb is critical as this is part of the appeal of eating honey in this way. There are several options to present and sell your honeycomb:

Cut-out presentation is the most common form of honeycomb consumption.

A frame, generally a shallow, is placed in a very active hive when a nectar flow occurs. The frame will have a thin strip of foundation at the top and may be wired. It is placed above a queen excluder, preferably between two fully drawn-out frames. This will encourage the bees to produce straight comb. It is critical to place this frame above the queen excluder so that no brood is ever produced in the comb as brood will leave some unpalatable residue. The frame is collected when it is fully capped with honey. If wire is present, the wire is cut at the side and pulled through. A sharp knife is used to cut the comb to the size of a square container (these can be purchased specifically for this purpose, or you can use Tupperware-type kitchen containers if for personal use). Make sure to minimize contacting the comb while manipulating it into the containers. Immediately freeze the container for at least three days to kill small hive beetle or wax moth eggs. Once frozen and thawed, it will store at room temperature for an extended time. However, depending on the source of honey it may crystallise in the comb.

Above: Chariot honeycomb frames are a popular method of obtaining comb honey for the table.

Right: Chariot squares with comb starting to be built.

In frame presentation of honeycomb is generally done using new shallow frames that can be presented in a specially adapted holder displayed on a breakfast table or buffet. Guests use a knife to scrape/cut some comb off and spread it immediately on bread. Some honey will flow down and is captured in the display device. This is ideal for upmarket hotels, B&Bs or buffet restaurants. To produce this type of presentation, follow the procedure above, but freeze and thaw the entire frame, ensuring the comb remains intact. Use kitchen freezer bags or plastic garbage bags to hold the frame for storage.

Honeycomb cassettes allow the beekeeper to produce (and market) honeycomb untouched by human hands. Indeed, these plastic inserts fit into special, standard-sized frames and have a honeycomb pattern embossed at the bottom. This will encourage the bees to build comb directly in the cassettes (especially if molten wax has been painted on them), fill it with nectar, and transform the nectar into honey. Each cassette produces a single side of the honeycomb. The beekeeper can collect the cassettes, place a lid on top of the cassette, and freeze and thaw the entire cassette prior to storage to kill any larvae. This system requires extremely strong hives and nectar flows, but when finetuned, it results in amazing examples of comb honey that can be sold at a premium. In a variation on this theme, there are also inserts made from thin pieces of wood called 'Chariot Honeycomb Frames'.

How do I produce wax, and what can I do with it?

Problem: I collect a lot of wax from discarded foundation and want to clean and use it.

Solution

Cappings left after extracting or comb cut-outs contain many impurities: pollen and honey are present, as are silk cocoons and other remains left by the pupae that need to be filtered out together with pieces of dead bee. The first stage in cleaning the wax is to melt it and filter it through an old cloth sheet, pantyhose, kitchen paper towel or piece of hessian sacking. The two most common methods hobbyists use to clean wax are the saucepan and solar beeswax extractor. The saucepan method is described here. An alternative is to build a solar extractor; details can be found in the book *The Australian Beekeeping Manual.*

SAUCEPAN METHOD

You can use a large saucepan from your kitchen if it is no longer used or find one at your local charity shop. It would be best if you dedicated a saucepan for this purpose as, once used, it will be challenging to clean it so it can be used again for its original purpose. Choose a size that comfortably fits all the wax you have for melting. Fill this about a quarter full of water and raise the temperature of the water on a stove, slowly adding the cappings until they have all melted. The resulting melted cappings are messy and contain unwanted material that will float on top. The beekeeper may need to negotiate the use of the family kitchen for this method and make sure any resulting mess or spilled capping material is thoroughly cleaned up. For those who can't negotiate the use of the family kitchen, it is also possible to use a second-hand electric frying pan, which can be plugged into an outlet in a more acceptable location.

While cappings are being melted, remain close to and regularly stir the wax to monitor its temperature. If the process is interrupted and the stove is left unattended, it should be turned off until the beekeeper can return and complete the operation. Indeed, overheating wax will invariably lead to a nasty fire once the wax temperature reaches its relatively low flashpoint at 205°C (400°F).

Before melting the wax, prepare a large bucket by adding about five centimetres (two inches) of water for the wax to float on while it hardens. Tie some cloth, kitchen paper towel, pantyhose or hessian sacking around the top to act as a filter and set it up outside the house. Make sure that there is a concave pocket in the cloth or sacking to catch the melted wax so that it will not overflow onto the ground beside the bucket. Next, carefully pour the wax into the cloth on the bucket. The melted wax will flow through the

Wax is melted in a large saucepan. Filtering molten wax through sackcloth.

cloth sieve, leaving the fibrous black mess left when cappings and comb are melted and filtered. The unfiltered foundation left in the cloth pocket or depression in the material is often called slum-gum and consists mainly of brood comb as it contains the silk cocoon, defecations and shed skin left behind by pupae as they develop and grow.

When all the wax has been filtered, and the wax both inside the bucket and on top of the cloth has hardened, remove the fabric and dispose of it in the bin or the compost. My experience is that although professional beekeepers have the equipment to extract further wax from slum-gum, this is very difficult for a hobby beekeeper, and a lot of time can be wasted extracting very little additional wax.

When the wax inside the bucket has hardened, it can be removed. If you turn the block wax over, you will notice some slum-gum has passed through the cloth and congealed to the underside of the block. Cut this off with a hive tool or sharp knife, and the remaining wax is usually sufficiently clean to swap for milled foundation from your local supplier.

Since beeswax will be discoloured by pots and containers made of iron, brass, zinc and copper, it is a good practice to melt wax only in containers made of aluminium or stainless steel.

If you plan to make candles or furniture polish from the wax, a further level of filtering or rendering will be required, and the process will need to be repeated. If the wax is to be used for cosmetic purposes, such as making face or hand cream, the process will need to be repeated yet again to obtain a higher-quality, pure beeswax.

How do I collect pollen, and what can I do with it?

Problem: I want to collect pollen to feed my bees during winter and make healthy drinks for the family.

Solution

Harvesting pollen is a good spring and summer activity when flowers bloom. Attaching a pollen trap to your hive is straightforward; you can collect a lot of pollen daily. A pollen trap comprises a perforated front plate and a collection box underneath. When the bees force their way through the perforations, the pollen is caught on the side of the perforations and removed from the bees' legs, falling into the collection box. In practice, since collecting pollen restricts how much pollen the bees can store and use to feed their brood, only use a pollen trap every other day. On the days that pollen is not collected, leave the entrance to the trap open so that the

1. Pollen traps can be attached to the entrance of a hive to collect pollen.
2. Pollen traps are easy to install and use.
3. Forager returning with pollen on her back leg.

returning foragers can enter and leave the hive freely. Pollen absorbs water and will quickly grow mildew if not harvested every evening. Once harvested, pollen is placed in a plastic bag and stored in the refrigerator or freezer.

Pollen can be fed back to the colony when there is a shortage of flowering plants. As a healthy food, pollen made into a milkshake or added to breakfast cereal makes a delicious meal.

How do I collect propolis, and what can I do with it?

Problem: I want to collect propolis from my hive and wonder what I can do with it.

Solution

Propolis is a black or dark brown resin-like substance bees collect from tree buds, sap flows or other botanical sources. Due to its antibacterial properties, bees use it to seal gaps in their hives and distribute it around the hive to protect against pathogens.

Propolis is a sticky substance and may be difficult to collect. For small quantities, use your hive tool to scrape it off the inside wall of the hive or the sides of frames. For larger quantities, a plastic propolis mat may be placed above the top super that workers will coat with propolis. A propolis mat is similar to a plastic queen excluder because it has a mesh construction that workers will try to block. For best results, when using the propolis mat, slightly open the top of the hive, allowing some light to enter. This will encourage the workers to deposit more propolis to seal the mat. Since a propolis mat is made of plastic, the best way to remove propolis is to roll up the mat and place it in a fridge. When the propolis is cold, it will break and can be removed from the mat.

When removed from the mat, propolis is a sticky, solid substance that cannot be used directly. It needs to be dissolved in alcohol before filtering out impurities.

Left: Propolis, after collection, is cleaned of impurities by dissolving in alcohol and filtering.

Below: Propolis in a hive can easily be collected.

Propolis has been used in various ways:

- **It's often taken as a supplement for its potential immune-boosting properties. Some people use it to help with colds, flu or other infections, although more research is needed to confirm its effectiveness.**
- **Propolis is used in skincare products for its antimicrobial and anti-inflammatory properties. It may help with conditions like acne or eczema.**
- **Its antiseptic properties make it helpful in treating minor cuts and burns.**
- **Propolis can be found in some toothpaste and mouthwash products due to its antibacterial properties that may help with oral hygiene.**
- **Some people use propolis to support gut health and manage digestive issues despite limited scientific evidence.**

CHAPTER 18

Varroa Mites

Varroa mite in an uncapped brood cell.

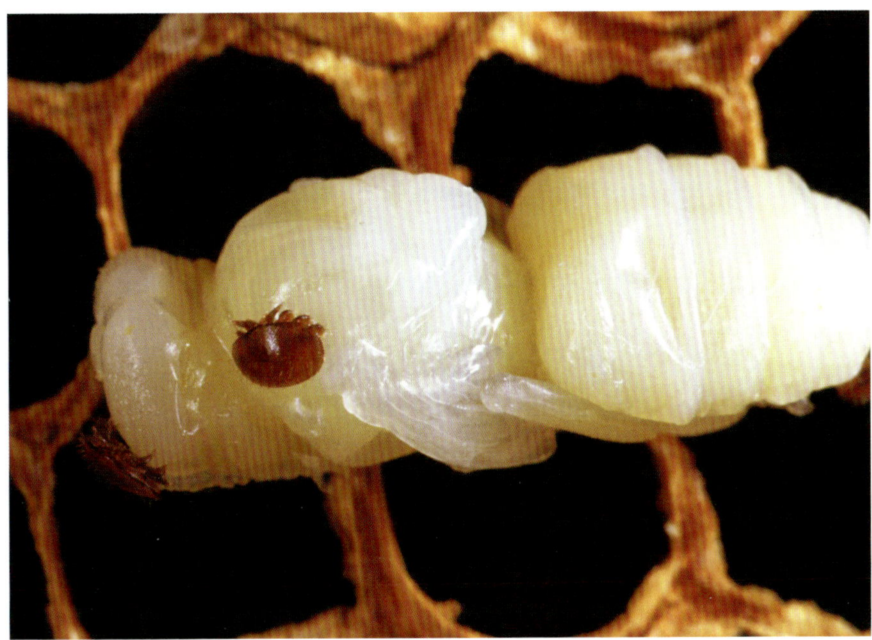

Varroa mite on a pupa.

What is Parasitic Mite Syndrome, PMS?

Problem: **I want to know about bee infestation with *Varroa destructor* leading to Parasitic Mite Syndrome, PMS.**

Solution

Parasitic Mite Syndrome (PMS) is caused by the parasitic mite *Varroa destructor*. These mites attach to bees and feed on their 'fat bodies', an organ similar to a liver in mammals. This weakens the bees and the *Varroa* mite transmits various viruses, which affect the hive and if left untreated will most likely kill the colony.

The symptoms of PMS in bee colonies include:

- **Infested colonies may show reduced population and less overall vigour.**
- **Bees that survive the mite infestation often exhibit physical deformities, such as deformed wings and legs, usually caused by the Deformed Wing Virus.**
- **Mites can negatively affect brood (larvae and pupae), leading to irregular brood patterns and increased brood mortality.**
- **The stress caused by mites can make bees more susceptible to other diseases and infections.**

Effective management of *Varroa* mites is crucial for the health and productivity of bee colonies. Beekeepers should use integrated pest management (IPM) strategies, including chemical treatments, mechanical methods and cultural practices to control mite populations and minimize their impact on bee colonies. Left untreated, *Varroa* mite infestation almost always leads to the hive's demise.

Above: *Varroa* mites are small but can be seen unaided by a beekeeper.

Right: Other mites often live in a hive that may be mistaken for *Varroa*. Top left, adult *Braula* fly; top right, adult male *Varroa*; bottom, *Tropilaelaps*, left, *Melittiphis*.

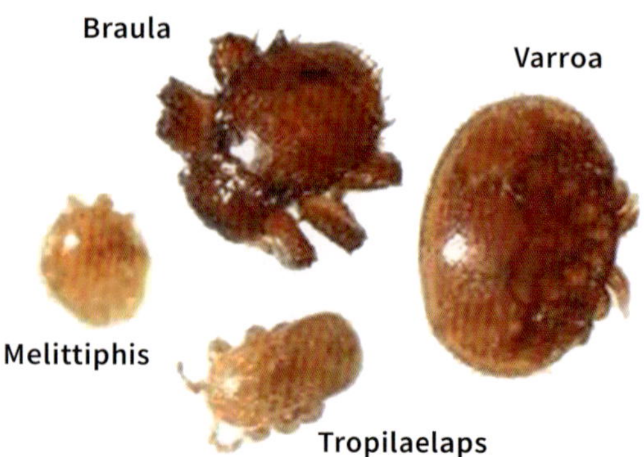

What should I do when my hive is infested with *Varroa*?

Problem: *Varroa* mites infest my hive. Urgent action is needed to manage this infestation.

Solution

Varroa mite infestation is the biggest challenge facing beekeepers globally. Managing this mite is not a one-time task but a continuous and complex process that requires ongoing learning and improvement, particularly as more and more resistance to current treatments is likely to emerge. This emphasis on continuous learning and improvement is not to overwhelm but to empower you as a beekeeper.

The *Varroa* mite spends most of its life inside capped pupal cells, living off the fat bodies of pupae. *Varroa*, on its own, is a manageable honey bee pest that causes some health problems. However, the problem is that while using the pupa for food, it injects harmful viruses, the most serious of which is the Deformed Wing Virus (DWV). This virus can lead to serious health consequences and the eventual death of the colony in one to two years if left untreated. There are no treatments for viral infections, and the beekeeper must keep the *Varroa* infestation below acceptable levels for the colony to survive. This underlines the urgency of managing *Varroa* mite infestation.

Estimating mite numbers

An essential part of the management process is regularly inspecting the colony to estimate the number of mites per 100 bees. The acceptable number of mites varies by region and season but is usually between about two and seven mites per 100 bees. This number is dependent on the prevalence of viruses transmitted by the mite. For example, during the 2022 *Varroa* outbreak in Australia, where common *Varroa*-transmitted viruses have not yet been detected, mite counts of several hundred per 100 bees were detected in hives, with minimal impact on these hives. This high mite count would be impossible in most parts of the world as the colony would have died long before then. If the number of mites is below the threshold, no treatment is needed until the next count. However, if the number of mites per 100 bees exceeds the threshold, immediate treatment with an approved miticide or organic chemical is not just recommended but necessary. This proactive approach is crucial for the health and survival of your colony.

The most popular tests for *Varroa* mites are presented below (see 'How do I perform a sugar shake to test for *Varroa* mites?', 'How do I perform an alcohol wash test for *Varroa* mites?', 'How to perform a sticky mat test for *Varroa* mites?' and 'How to perform a drone uncapping test for *Varroa* mites?').

Management

Varroa management is a time-consuming and complex process that can potentially harm your colony if performed incorrectly. An integrated pest management (IPM) approach is essential, and a combination of management methods should be used to minimize the use of chemicals and the development of resistance to treatment.

Using the IPM approach outlined in the diagram opposite, a mixed approach of rotating management methods provides the optimum strategy for combatting *Varroa*.

Cultural

Cultural approaches include basic beekeeping practices that should be performed even if there is currently no *Varroa* in the hive. These include using *Varroa*-resistant bees and regularly breaking the egg-laying cycle of the queen by caging her for approximately three weeks. They also include keeping strong colonies and merging weak colonies (see 'I have two weak colonies; how can I merge them?').

Mechanical

While cultural controls should be used whether the colony is infested or not, mechanical controls are helpful if a *Varroa* mite infestation is present

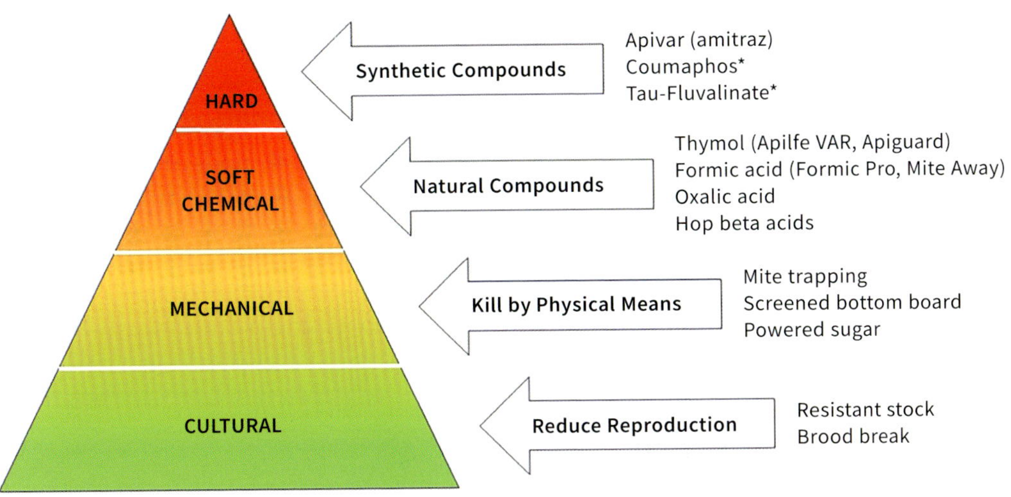

* The use of Coumaphos and tau-fluvalinate are not recommended in some countries since *Varroa* has developed resistance to these chemicals

An integrated pest management plan (IPM), must be designed and implemented for your apiary if pathogens are to be kept to a minimum, development of resistance to miticides minimized and to minimize use of chemicals in and around the hive. See 'I want to implement an integrated pest management (IPM) strategy for my apiary.'

or likely. Mechanical controls include screened bottom boards, drone brood removal and powdered sugar dusting. Since these do not involve the use of chemicals, they are part of a good IPM strategy to ensure the long-term viability of the bee population.

Chemical

Colony death may result during winter if the *Varroa* population is uncontrolled during spring and summer. In late spring, soft chemicals may reduce the mite population to an acceptable level.

Varroa mite reproduction in spring and summer often leads to a large population in late autumn and winter. Suppose an infestation level is reached where mite numbers may harm your colony. In that case, you will have the best overwintering success if a chemical miticide is applied before the production of the winter bees. In an IPM system, soft chemicals are used before moving to hard chemicals. Chemicals should only be used if testing shows the number of mites per 100 bees is above the safe level for the colony, which will vary depending on your location.

Organic acids such as formic and oxalic acids, essential oils such as thymol, and hop beta acids (from the brewing industry) occur naturally. Synthetic miticides are pesticides developed to control *Varroa* mites and should be used as a last resort due to their potential to harm your colony in the short term and the bee population in the long term. The development of miticide resistance would harm the bee population in the long term. Without resistance, synthetic miticides, including Amitraz and Coumaphos, kill up to 95 per cent of mites in a colony.

Summary

Many techniques are available to manage *Varroa*, and the topic is too complex to discuss adequately in this book. Readers need to talk with local beekeepers, a local beekeeping supply store or a government apiary inspector to determine what is effective in their region. Each technique has advantages and disadvantages, so a good IPM strategy tailored to your situation will ensure the optimum well-being of your bees.

A mesh baseboard provides some protection against *Varroa*, but it is insufficient to manage the infestation alone.

Why does the *Varroa* mite count change predictably over the seasons?

Problem: I want to know why the *Varroa* mite count changes over the seasons and how I treat my hives.

Solution

Varroa infestation can be estimated using several methods, and the counts obtained can be fundamentally different. Indeed, some methods, such as alcohol wash and sugar shake, estimate the number of mites per 100 bees, while the sticky mat method estimates per hive (number of mite drops per time unit per hive). These are fundamentally different because the mite per 100 bees count highly depends on the colony's size. Hence, these two values can't be compared unless the colony size is considered.

Varroa mites use capped brood to replicate. Hence, the mite-drop count trails the amount of brood in the hive, allowing us to estimate when most mites will be found without treatment. For example, in temperate climates, the hives produce a lot of brood during the spring build-up. This includes many drones, and drone brood is more efficient than worker brood for mite replication due to its longer capped brood period. Therefore, it is expected and experimentally confirmed that the highest mite-drop numbers are in summer. As summer turns into autumn, the amount of brood diminishes as the hive peaks and prepares for winter. The reduced brood and absence of drone brood will also result in lower mite-drop counts. In many parts of the world, during winter, there might be a period without any brood, and this brood interruption will dramatically affect mite drop counts. Hence, we expect the lowest mite-drop counts at the end of winter.

To convert these numbers to the mite count per bee we need to consider the size of the hive. At the start of spring, when the hive comes out of winter, we expect the lowest bee numbers. This number builds up rapidly in early spring. During that time, both bee and mite numbers increase, with the bee numbers tending to be ahead of the mite numbers. Thus, mite

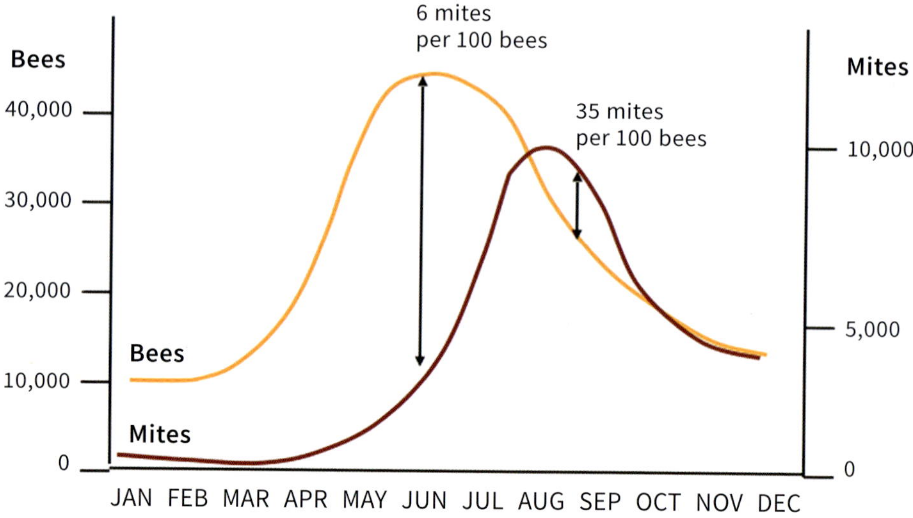

Simplified bee and mite population growth curves for a temperate climate. The mite growth curve lags behind the mite growth curve. Note how the number of mites per 100 bees significantly increases in autumn. A colony is unlikely to survive this high autumn infestation rate — diagram modified from www.scientificbeekeeping.com.

counts per bee will not increase dramatically, at least not initially. However, mite numbers soon catch up, and mite numbers per bee will increase by the end of spring and summer. In autumn, the number of mites is rising (at least initially) while bee numbers decrease. Hence, the number of mites per bee will increase rapidly and dramatically. As the bees start to cluster for winter, the number of bees stabilises, but because the amount of brood diminishes (sometimes to zero), the mites can no longer reproduce, and the number of mites per bee decreases.

Therefore, spring and autumn are the most critical periods for mite control. Early spring treatment will help prevent the number of mites from increasing too rapidly during hive build-up and drone production, which would affect the hive later in the year. Another treatment in early autumn will help prevent the bees from being overwhelmed by a large number of mites for a diminishing bee population. There will be many local variations on this general theme, and beekeepers should consult with local clubs, beekeeping equipment suppliers and experienced beekeepers.

How do I perform a sticky mat test for *Varroa* mites?

Problem: I need to check the number of *Varroa* mites in my colony using the most sensitive method available.

Solution

The sticky mat, drone uncapping, sugar shake and alcohol wash methods are the main ways to detect and estimate the number of *Varroa* in a colony. Each has advantages and disadvantages, which are discussed in turn.

A sticky mat test for *Varroa* mites is a practical, non-intrusive method to estimate the level of *Varroa* mite infestation in a beehive. Here's how you can perform it:

Materials needed:

- Sticky board (or a piece of white paper or cardboard).
- Mesh screen (8-mesh hardware cloth, big enough to allow mites to fall through but prevent bees from falling through). 8-mesh corresponds to 3mm or 1/8th inch mesh.
- Vaseline, cooking oil spray or sticky adhesive.
- Gloves (optional).

Steps to perform the test:

1. Prepare the sticky board:

- Take the sticky board and cover one side of the board with a thin layer of a sticky substance like Vaseline, cooking oil spray or a commercial adhesive.
- If you're using paper or cardboard, you can apply the sticky substance directly to it.

2. Place the sticky board in the hive:

- Insert the sticky board under the hive's screened bottom board. If your hive doesn't have a screened bottom board,

Left: A sticky mat is inserted into a hive's base.

Below: A sticky mat is placed on a bottom board under a mesh to keep bees away. After one or, at most, two days, depending on the level of infestation, many mites can be seen on the mat. Also on the mat are debris, parts of bees and pollen.

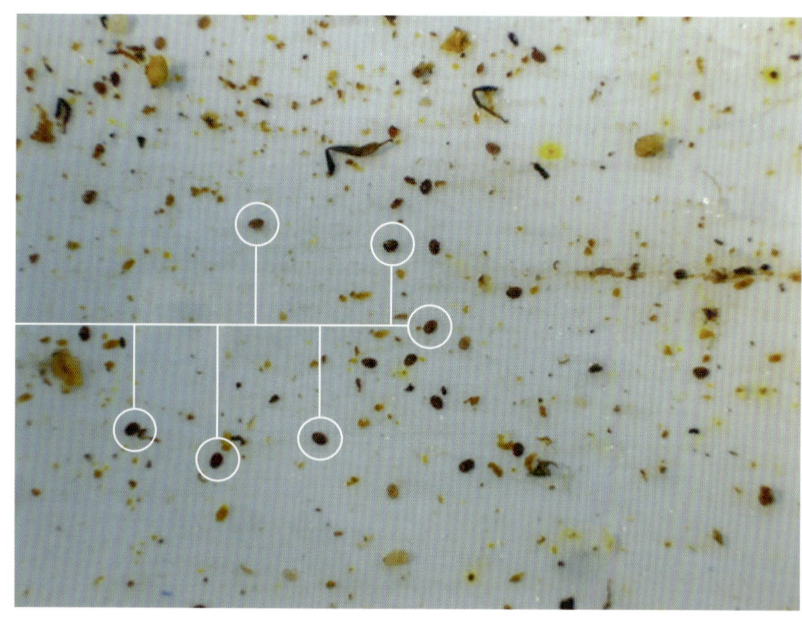

you'll need to place a mesh screen over the sticky board. This prevents bees from directly contacting the sticky surface while allowing the mites to fall through.

3. **Leave the board in place:**
    - Leave the sticky board inside the hive for 24 to 72 hours. Leaving the sticky board in for extended periods (72 hours) provides a more accurate count.
4. **Remove the sticky board:**
    - After the desired time period, remove the sticky board from the hive carefully.
5. **Count the *Varroa* mites:**
    - Inspect the sticky board and count the number of *Varroa* mites that have fallen onto it. Use a magnifying glass if needed to spot them more easily.
6. **Calculate the mite fall rate:**
    - Divide the number of mites by the days the sticky board was in place to get an average daily mite drop. This gives you an estimate of the mite load in the colony.

Interpretation of Results:

- Ask local beekeepers or an apiary inspector the maximum number of mites detected on a sticky board before management is needed.
- Fewer than ten mites/day may indicate a low infestation.
- 10–50 mites/day suggests a moderate infestation, and you should monitor the colony more frequently.
- A high infestation is considered when there are more than 50 mites per day and urgent treatment is needed.

This method passively monitors mite populations without directly disturbing the bees. This is especially true when using a screened bottom board with a sliding insert for the sticky cardboard.

How do I perform a sugar shake test for *Varroa* mites?

Problem: Using the sugar shake method, I must estimate the number of *Varroa* mites per 100 bees.

Solution

Sugar shake monitoring is probably the most widely used *Varroa* mite surveillance tool. If performed correctly, vigorous shaking is required, resulting in the death or injury of many bees.

Materials needed:

- Container with a lid adapted with a metal screen.
- Container or dish with water (preferably white colour to facilitate mite count).
- Cardboard, bowl or newspaper to collect the bees shaken from the frames.
- Icing sugar.

Steps to perform the test:

1. Collect nurse bees:

- Remove a brood frame from the hive, not a frame of honey, and check the queen is not present.
- Brush or shake the bees into a bowl or onto a sheet of newspaper/cardboard and wait about a minute for the older forager bees to fly off. This step will substantially increase *Varroa's* probability of being detected in an infested hive.
- Funnel or scoop about one cup of bees into a jar. Add approximately one heaped tablespoon of powdered (preferably sifted) icing sugar to a jar. Secure the mesh lid onto the jar.

2. **Detach mites from the bees:**
    - Vigorously shake for one minute to cover the bees in sugar and dislodge the mites from the bees. Sugar-covered bees must be vigorously shaken so that some may die. Otherwise, mites may not be dislodged, giving a false negative test. Your arm will get tired, so take a rest after 20 seconds.
    - Set the jar down and wait three to five minutes. Rushing the process increases the risk that an infestation will not be detected.
3. **Shake the sugar and mites out of the container:**
    - Invert the jar and vigorously shake it like a saltshaker over a pan of water. It is better to shake the jar over water rather than a clean plate since mites float on top and are more easily identified. The sugar will also dissolve in the water, making mite detection easier.
    - Add another tablespoon of icing sugar to the jar, shake and roll the bees again for 30+ seconds, and repeat the previous steps to improve the test's sensitivity.
4. **Inspect for mites and release the bees:**
    - Inspect the pan of water to see if any mites are floating on the surface. There may also be dirt floating, so glasses or a magnifying glass may be needed.
    - Release bees tested at the top of their colony or the colony entrance.

Since we are estimating the number of phoretic (i.e., attached to adult bees) mites per 100 adult bees, if we sampled 300 bees and found nine mites, the infestation rate would be 9 per 300 = 3 mites per 100 bees. In North America, this level of infestation, depending on the season, may require immediate treatment of the colony with a miticide. In other parts of the world, the number may differ depending on the prevalence of viruses transmitted by *Varroa*. The more viruses present, the lower the

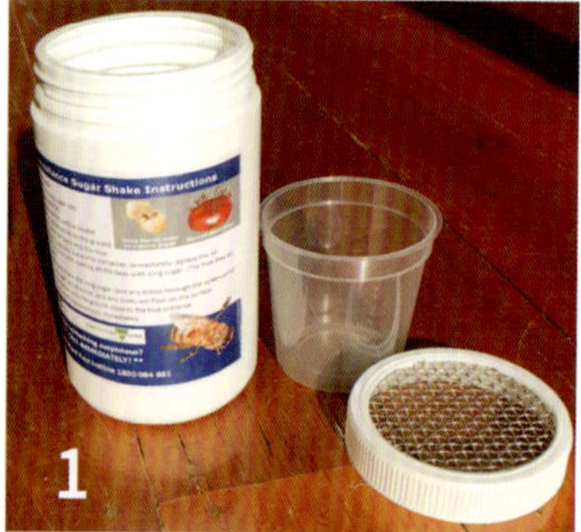

threshold before requiring treatment.

During the initial stages of an incursion, mite counts in a hive are likely low. In this case, the jar containing sugar-coated bees **must be shaken vigorously for up to a minute** to maximize the probability that any mites will be dislodged and detected. Many descriptions of the sugar shake technique say to roll the jar of bees gently. Gently rolling the jar is unlikely to dislodge any mites. Although beekeepers are reluctant to kill bees during testing, the jar must be sufficiently agitated that some bees under test will likely be destroyed or damaged. Without vigorous shaking, an accurate estimate of the infestation level in a colony cannot be made. Count the number of bees being tested if you have only recently started monitoring *Varroa* levels to increase testing accuracy. The number does not have to be precisely 300, although the more bees in the sample, the more accurate the test will be. Adjust the number of bees in the formula to calculate mite load per 100 bees.

1. Equipment needed for a sugar shake test.

2. Adding finely ground sugar, icing sugar, to the sugar shake jar.

3. A cup full of bees, about 300 bees, is added to the sugar in the jar.

4. After vigorously shaking, the jar is inverted and the mites and sugar are shaken out. The mites are best shaken onto a plate or bowl of water where the sugar will dissolve.

How do I perform an alcohol wash test for *Varroa*?

Problem: I need to estimate the number of *Varroa* mites in my colony per 100 bees using an alcohol wash.

Solution

Although the alcohol wash method is often referred to as the gold standard for estimating *Varroa* numbers, there is no evidence that either the sugar shake or the alcohol wash produces more accurate results. Both methods have their limitations with accuracy, so choose whichever method is most convenient for you to use.

Most beekeepers know of the alcohol wash detection method, but many may not know that dishwashing liquid may be used instead of alcohol with the same accuracy. In the following description of the alcohol wash, the same steps must be taken if dishwashing liquid is used. In North America, Dawn washing liquid may be used; in Europe and Australia, the Fairy brand is suitable. Using dishwashing liquid instead of alcohol is particularly useful in areas prone to bush or wildland fires. Indeed, an arid environment combined with a hot smoker and alcohol can easily result in a catastrophic fire that would destroy not only the apiary but also vast areas of natural habitats.

Materials needed:

- Transparent container with a screened insert. There are many commercially available kits.
- Alcohol or dishwashing liquid and water.

Steps to perform the test:

1. Collect nurse bees:

- Remove a brood frame from the hive, not a frame of honey, and check the queen is not present.

- Place 300 bees in a jar and cover it with isopropyl alcohol, methylated spirits or water with some dish-washing liquid. The dishwashing liquid must be diluted to about ¼ to ½ tablespoon per litre of water (one to two tablespoon/gallon).

2. Detach mites from the bees:

- Shake vigorously for one minute to dislodge the mites from the bees. Your arm will get tired, so take a rest after 20 seconds.
- To ensure consistency between tests or colonies, shake each sample with the same vigour and for the same amount of time for each test.

3. Count the mites:

- After shaking, check the bottom of the jar below the screened insert or empty the liquid into a strainer, with the alcohol falling into a container.
- Add more alcohol to the jar of bees and vigorously shake again. This increases the accuracy of the test.
- Empty the alcohol into the strainer and let the liquid collect in a container with the previous liquid.
- The strainer must allow mites to pass through without allowing bees to pass through.
- Count the number of mites in the container.
- Discard the dead bees well away from the hive. If using alcohol, keep it away from any open flame.

The same method is used to estimate the infestation rate for sugar shakes. If 300 bees are tested and nine mites are dislodged, the infestation rate is 9 per 300 = 3 mites per 100 bees.

1. Equipment needed for an alcohol wash. Rubbing alcohol from a hardware store is suitable.

2. Three hundred bees and alcohol are sealed in the jar, which is shaken vigorously.

3. The sieve containing dead bees is removed from the jar.

4. Mites are left floating in the alcohol remaining in the jar.

How do I perform a drone uncapping test for *Varroa*?

Problem: I need to check for the presence of *Varroa* in my hive using the drone uncapping test.

Solution

Drone uncapping is a very rapid way of assessing the infestation of a hive, although it may be less quantitative compared to other methods. The technique is based on the fact that *Varroa* mites prefer drone brood to worker brood. Drone brood takes longer to emerge than worker brood, giving the mite more time to reproduce. Over time, *Varroa* mite has evolved to seek out drone brood preferentially. Also, drone brood is less valuable to beekeepers (unless they are seeking to spread a particularly good genetic stock) because drones do not collect nectar or participate in other hive maintenance activities. Therefore, many beekeepers are okay with losing some drones.

Materials needed:

- An uncapping fork.
- Alcohol or dishwashing liquid and water.

Steps to perform the test:

1. Identify drone brood:

- Check the brood box and identify capped drone brood. Drone brood is produced in larger cells, and the capping protrudes from the frame's plane.
- Most beekeeping suppliers and online stores sell plastic frames (usually green in colour) embossed with a larger, drone-sized cell pattern. Worker bees building comb on these frames make larger cells in which the queen will lay unfertilized drone eggs, which will become drones. Therefore, if you use a (usually) green frame designed to favour drone production, the capped brood on that frame should be drone brood.

Above: Uncapping drone comb. Drone cells stick out from the surface of the brood comb and are bullet-shaped. Worker brood comb is flat.

Right: Mites on uncapped drone pupae.

2. Uncap the drone brood and look for mites:

- Using an uncapping fork, remove the capping of the drone brood, pulling out the pupae.
- Inspect the pupae, looking for mites (small brown dots), which are now easily spotted against the white colour of the drone pupae. A total of 100 or so drone pupae need to be inspected.

See also 'How can I manage *Varroa* using drone brood removal?'

How can I manage *Varroa* using drone brood removal?

Problem: **I want to use drone brood removal to help manage a *Varroa* infestation.**

Solution

Varroa mites prefer drone brood over worker brood and are eight to ten times more likely to be found in drone brood than worker brood. Although drone brood removal is insufficient to manage *Varroa* on its own, it is a valuable addition to the tools available to a beekeeper to control an infestation. Since drone brood removal techniques are fundamentally different from chemical treatments, there is no possibility that they would help with resistance development. Hence, they are an ideal tool for an integrated pest management system (IPM).

Most beekeeping suppliers and online stores sell plastic frames or foundation (usually green) embossed with a larger, drone-sized cell pattern. Worker bees building comb on these frames make larger cells to rear drone larvae and pupae. As a result, if the drone comb is placed in the brood box, the queen will lay unfertilized drone eggs in the larger drone comb. When the cells have been capped, most of the *Varroa* in the hive will be found in this drone comb. The beekeeper removes the plastic drone frame and destroys the comb by freezing the frame, killing the bee-pupae and the *Varroa* mite, and significantly reducing the number of mites in the hive. The drone frame is stripped of the comb, coated with a layer of wax and then returned to the hive. The removal of drones does not affect the ability of the hive to function but would limit the distribution of the hive's genetics. Indeed, drones' primary function is to fertilize a queen from other colonies (see 'How can I recognize different castes of bees?').

The steps to be taken are:

1. In late April or May in the UK and early spring elsewhere, inspect the brood chamber to confirm that bee numbers in the colony are increasing.

MANAGING *VARROA* USING DRONE BROOD REMOVAL

Left: Green plastic drone frames can be obtained online or at a beekeeping supply store.

Below: Replacing a standard frame with a drone frame.

2. Remove an empty frame or frame containing mostly stores from the brood box, next to frames containing worker brood, but not amongst brood frames, and insert a drone frame.
3. About three weeks later, open the brood box and confirm that capped drone brood cells are present. Do not leave the inspection longer than 24 days since drones take 24 days from egg to hatching as adult drones, releasing any mites in the capped cells. If the frame contains a large percentage of capped brood, remove the frame and destroy the cells. You can take this opportunity to uncap some of the capped brood to assess the presence of mites in the capped drone brood. Burn the foundation or place the frame in a freezer overnight to kill the mites.
4. Either return the drone frame to the hive, replace it with a new drone frame or replace it with a frame of worker foundation.

See also 'How do I perform a drone uncapping test for *Varroa*?'

How can I manage *Varroa* by caging the queen?

Problem: To help manage *Varroa*, I want to use a queen cage to limit her ability to produce brood and interrupt the mites' life cycle.

Solution

In a typical hive, the queen lays up to 1500 eggs a day, so there is usually a reliable supply of uncapped larval cells about to be capped. Mites in the hive use this supply of brood cells that are about to be capped to make their home, breed and increase their numbers. The principle of queen caging is to stop the queen from laying eggs by restricting her movements while still in the hive, where she is looked after by nurse bees. Since the queen is unable to lay eggs, the supply of uncapped brood cells about to be capped is stopped. This breaks the mites' life cycle since they do not have larval cells that are about to be capped. This has the advantage that most mites are outside of capped cells, where they can more easily be killed with miticides. Queen cages come in various designs and sizes — small cages to hold the queen for up to a week — and larger cages, sometimes including foundation, that can keep the queen for several weeks. Caging a queen should be performed early in the season, allowing the colony to regain strength before autumn.

The steps to be taken are:

1. Holding the queen in a small cage.

1. Place the queen in the cage through a door on the cage.
2. Place the cage next to brood cells in the hive.
3. Leave the queen in the cage until all the uncapped brood cells have been capped.
4. Release the queen from the cage.
5. This may be an excellent time to treat the hive with Oxalic acid, killing mites not in capped cells (see 'How do it treat my hives for *Varroa* mites with different acids?').

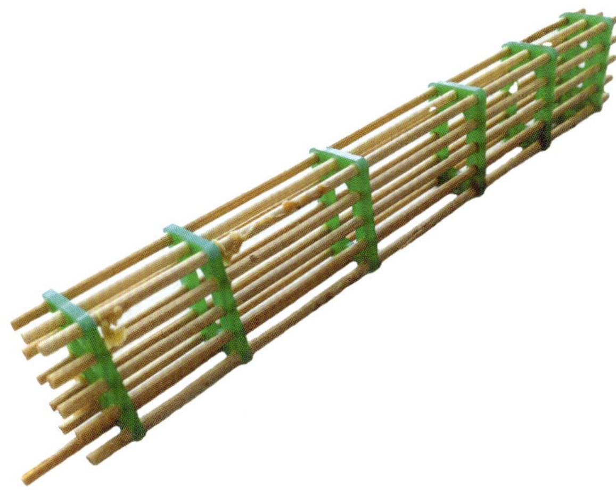

A cage to hold the queen for up to about 28 days: the queen and cage can be left in the hive where the workers will feed her. She will be unable to lay eggs while in the cage, breaking the life cycle of the *Varroa* mites. *Tropilaelaps* can also be managed using a queen cage.

2. Holding the queen on a frame.

1. Using a frame of drawn comb, cut a hole in the corner of the foundation so the queen can quickly move between both sides.
2. Remove an empty or brood-less frame from the brood box.
3. Place the drawn frame comb with the queen inside the cage and into the centre of the brood box amongst brood cells.
4. After twenty days, remove the frame and destroy any capped cells. Return the queen to the cage with a new replacement frame of drawn comb.
5. After 24 days, no brood should be left in the hive outside the cage. The queen can be released and allowed to resume laying eggs.

How can I manage *Varroa* by hive-splitting?

Problem: **I want to manage *Varroa* by splitting my colony and wonder how best to achieve this result.**

Solution

Managing *Varroa* using hive splitting is identical to the techniques described in 'How do I split a hive?' for Langstroth hives and 'How do I split a top bar hive?' for top bar hives. When a hive swarms, which splitting emulates, the old queen and brood are transferred to a new hive with frames of drawn comb. Some mites may be transferred with the swarm to the new hive. It will take about two weeks for eggs to be laid and larvae in uncapped cells to be ready to pupate and have their cells capped. This would be an excellent time to apply miticides since most of the mites will be phoretic and vulnerable, that is, living on the body of an adult bee.

The colony remaining in the old hive will be queen-less for a few days. If the hive were split because queen cells were found in the old hive, it would take about fourteen days for the new queen to go on a mating flight, lay eggs, and have the larvae ready to pupate and have their cells capped. Again, this would be an excellent time to apply miticides. If, instead, a new, purchased queen is to be placed in the old hive, you can delay inserting the queen and keep her in a cage slightly longer, further disrupting the capping of cells.

See 'How do I split a hive?'

What are organic and synthetic chemical treatments for *Varroa?*

Problem: Many types of chemicals are available to manage *Varroa*. The terms organic and synthetic chemicals are often used. What does this mean?

Solution

There are several chemical treatments for *Varroa*, including organic acids, essential oils and synthetic (man-made) miticides. An integrated pest management (IPM) plan would emphasise soft chemicals over hard chemicals since they do less harm to bees. The essential part of any plan would be rotating different *Varroa* control strategies including different chemicals, so resistance will take much longer to develop. Chemicals frequently used to control *Varroa* are listed in Table 6 opposite:

Organic acids and essential oils	Formic acid	Penetrates wax cap and kills reproducing mites. Temperature dependant. Do not use above 30°: queen mortality, brood loss. Low effectiveness below 10°.
	Oxalic acid	It does not penetrate cappings. It is more effective when there is less brood. Best used during winter or spring – low brood numbers. Apply using vapour or dribble. Should be used with other methods. Do not over-use – harms bees and decreases activity and longevity.
	Thymol	Most popular essential oil for *Varroa* mite control. Does not penetrate cappings. More effective when there is less brood. Efficacy of thymol treatment can be low. Combine with other treatment methods. May increase the aggression of colonies and induce robbing.
	Hop beta acids	Safe for use any time of the year, even during honey flows. The potassium salt of hop beta acids is derived from the hops plant. Does not penetrate cappings. More effective when there is less brood. Not affected by temperature. Not as effective as other soft chemicals.
Synthetic Pesticides Effective, kills up to 95 per cent of mites. Synthetic miticides should be used as a last resort as they harm bees.	Amitraz - Apivar®	The most popular synthetic miticide. Linked to increased bee mortality. Mites slowly developing resistance.
	Coumaphos - Checkmite+®	Not recommended in most regions due to the development of resistance.
	Fluvalinate - Apistan®	A synthetic pyrethroid. Often called tau-fluvalinate. Not recommended in most regions due to the development of resistance.

How do I treat my hives for *Varroa* mites with different acids?

Problem: The treatment of *Varroa* mites with different acids has advantages and disadvantages, which I want to explore to decide whether this approach to mite control is appropriate for my apiary.

Solution

Some organic acids can treat *Varroa* because they are more toxic to mites than bees. Hence, at specific concentrations, mites are affected more than the bees; thus, mite counts are dramatically reduced. Two acids are discussed below:

Oxalic acid

Oxalic acid is naturally present in plants but is toxic, including to humans. It can be delivered by different methods. We recommend vaporization and dribbling, while using an insect fogger has been reported to be less effective. Oxalic acid should be applied while wearing protective equipment, including gloves, goggles and respirators. It should not be applied when supers are present to avoid contaminating honey destined for human consumption.

Vaporization requires a special electric device to vaporize the oxalic acid crystals. The vaporizer is inserted in the hive's entrance, and the heat generated sublimates the oxalic acid, transforming it into a gas that will spread inside the colony. A total of 4 g of oxalic acid per hive is applied.

Dribbling works best when the colony is dormant and clustered. A warm sugar solution (1:1 sugar to water by weight) containing oxalic acid (35 g/L or 5 oz/gallon) is squirted between frames (5 ml between each frame) at a maximum dose of 50 ml per hive.

Oxalic acid is typically applied during the autumn or winter phase when the total amount of brood is reduced. Treatment is less effective during the build-up phase (spring) and the bee peak population (summer). This is because oxalic acid is most effective at removing phoretic mites (i.e., those on adult bees) and is ineffective at eliminating mites from capped brood.

Formic acid

Formic acid may be applied in strips or pads. Several brands can be purchased, and you should follow manufacturers' instructions. One well-known example is Formic Pro made by NOD Apiary Products Ltd. Optimal treatment with formic acid needs to occur within a tight outside temperature range (10–26 °C; 50–79 °F). Too high temperatures (above 29 °C; 85 °F) will result in excessive damage to the hive, while too low a temperature will not effectively reduce mite numbers. Treatment will last for seven days, and after that period, it is possible to put supers back on the colony to collect honey for human consumption. For treatment after a honey flow, remove the supers before treatment to avoid contaminating the honey. Some products may be approved for use during a honey flow as formic acid naturally occurs in honey. Check manufacturers' instructions.

Formic acid is corrosive and will affect skin, eyes and mucosal surfaces. Wear appropriate protective equipment as recommended by the manufacturer. Since formic acid is applied via impregnated strips, the amount of vapour is limited, and a respirator is not considered essential (in contrast to oxalic vaporization). Unlike oxalic acid, formic acid affects mites inside the capped brood. Treatment is recommended during the build-up phase (i.e., spring) when mite numbers are increasing and during the decrease phase in autumn to reduce mite numbers going into winter and minimize winter losses. Formic acid can also be dissolved in syrup and fed to the colony.

Thymol

Thymol, a natural compound in thyme oil, is the most popular essential oil for *Varroa* control. Thymol works as a miticide and can be part of an Integrated Pest Management (IPM) strategy for controlling *Varroa*. Thymol has both direct toxic effects on *Varroa* and a fumigant action, affecting mites outside of capped brood. It also has some antimicrobial properties, which may contribute to the general health of the hive by reducing pathogens like

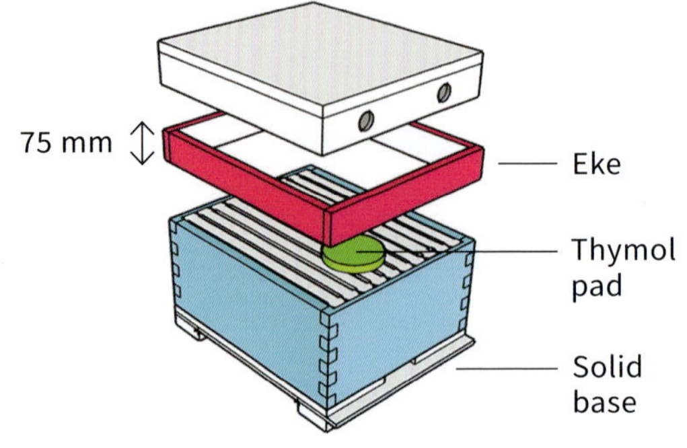

An eke is placed above the brood box, providing room for a pad of Thymol. Similar in design to a hive box, an eke is only 75 mm tall. Ekes are also helpful when feeding syrup or pollen inside the hive.

Top: Applying oxalic acid during summer using the drip method. Apply approximately five grams per hive.

Above: Using a vaporizer to apply oxalic acid. Apply approximately two grams per hive.

bacteria and fungi. Thymol works best in warmer temperatures (between 50°F and 85°F or 10°C to 30°C). When applied in warmer weather, it will evaporate, fumigating the mites. In colder weather, however, its effectiveness decreases.

Thymol can be applied either as a gel or by vaporization. It is often applied after the honey harvest when the colony's capped brood numbers are lower. This timing ensures fewer sealed brood cells where mites can hide since the oil will not penetrate the wax capping. Overuse or incorrect application can stress the bees or disrupt hive dynamics.

When used correctly, thymol can be an effective and relatively gentle option for managing *Varroa*. However, like all treatments, it should be part of a comprehensive IPM management plan that includes monitoring mite levels and using multiple control methods to ensure long-term effectiveness and colony health. Always follow the manufacturer's instructions for the specific thymol-based product you are using, and consider local beekeeping conditions when choosing your treatment strategy.

How do I treat my hives for *Varroa* mites with amitraz?

Problem: I want to treat *Varroa* in my hive using amitraz?

Solution

Amitraz, sold under the trade name Apivar©, is typically administered using strips placed inside the hive. These strips slowly release the active ingredient over weeks, ensuring prolonged mite exposure and enhancing the treatment's effectiveness. Commercial beekeepers have consistently used amitraz for two decades, and some are wondering whether their mites are finally starting to develop resistance to it, as they did rather rapidly to fluvalinate and coumaphos.

Amitraz is a contact miticide delivered when bees walk on the strip's surface. They pick up molecules of the active ingredient and then distribute them throughout the colony. Many beekeepers believe amitraz can only be applied using strips impregnated with the chemical due to its low vapour pressure. Recent experiments have shown that amitraz can be spread as a vapour.

Amitraz is harmful to bees and humans and should not be used when honey is present in a hive about to be harvested. Apivar® can be used to treat *Varroa* mites in a beehive twice a year, once in the spring and once in the autumn:

Spring: Use Apivar® two months before the honey season begins.

Autumn: Use Apivar® after the honey harvest.

Strips containing amitraz are inserted into a hive.

How do *Varroa* mites transmit viruses, and why does it matter?

Problem: *Varroa* mites are known to transmit viruses. I want to know what these viruses are, how they can affect the colony, and what can be done to help the colony survive.

Solution

In areas where *Varroa* mite has been present for many years, this mite will transmit viruses. These viruses, including the Deformed Wing Virus (DWV) and Black Queen Cell Virus (BQCV), are thought to be the primary cause of colony death. Unfortunately, there is no direct treatment for the viruses. Therefore, beekeepers need to minimize mite loads to prevent colony collapse. The number of mites a colony can tolerate depends on where you keep your bees. Check with local departments of agriculture (or equivalent), beekeeping clubs, beekeeping stores or experienced beekeepers.

Varroa mites transmit viruses by feeding on bees' fat bodies. In the process, the mite punctures the pupa's skin, injects fluid to digest the fat body and inadvertently infects the pupa with the virus. Since transmission occurs mainly through physical factors, *Varroa* mites can transmit many viruses, some more harmful than others. Deformed Wing Virus (DWV) can replicate in the *Varroa* mite without harming the mite; hence, transmission is amplified when *Varroa* is present.

If the mite is not infected with a virus, it can become so by feeding on an infected pupa.

Carniolan bee infected with Deformed Wing Virus.

Is my colony infected with a virus?

Problem: I want to know whether viruses infect my colony and what I can do.

Solution

Honey bees can be infected with over twenty viruses, some of which do not show any symptoms or appear to harm the colony. A recent study showed that many bees are infected with several viruses, including Deformed Wing Virus (DWV), Black Queen Cell Virus (BQCV), Sacbrood virus (SBV) and two Paralysis viruses. Deformed Wing Virus is associated with *Varroa* and is spread when the mite feeds on bee pupae.

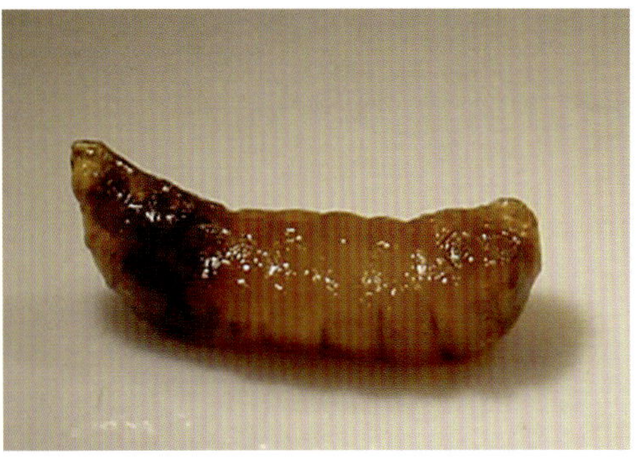

A larvae infected with Sacbrood virus. Infection with Sacbrood is often mistaken for infection with EFB.

Options for diagnosing viral infections are limited since only some viruses have symptoms that can be used to detect their presence, while others do not show symptoms. Options for dealing with viral infections are also limited since there is no specific treatment for viral infections in honey bees. However, beekeepers can take steps to minimize viral transmission and reduce exposure to other stressors, such as parasites, pesticides and nutritional deficiencies, by following a recognized integrated pest management (IPM) strategy. See the section 'I want to implement an integrated pest management strategy for my apiary.'

CHAPTER 19

Other Pests and Diseases

This chapter outlines how to identify and manage the most common pests and diseases. A more complete discussion is found in the book *Honey Bee Pests & Diseases*, which the authors recommend you read for further details. In many jurisdictions, the more serious of these diseases must be reported to local government authorities if they infect or infest your colonies. Check online or with government apiary or extension officers to confirm which need to be reported.

No beekeeper wants to see this in their hive.

A colony infested with chalkbrood.

How do I implement a comprehensive IPM plan for my apiary?

Problem: I want to minimize the use of chemicals to manage pathogens and use an integrated pest management (IPM) strategy.

Solution

Developing an Integrated Pest Management (IPM) strategy for your apiary involves implementing steps to ensure your bees' health and productivity while minimizing pesticide use. Factors to consider are as follows:

Identify Pest and Disease Risks

- Conduct regular inspections of your hives to identify any signs of pests or diseases.
- Educate yourself about common pests and diseases affecting your region's honey bees. These could include *Varroa* mites, wax moths, small hive beetles, American foulbrood, European foulbrood, etc.

- Keep records of pest and disease occurrences and note any patterns over time.

Establish action thresholds

- Decide at what point pest or disease levels require action. This could be based on quantitative measures (e.g., the number of mites per hundred bees) or qualitative observations (e.g., visible damage to the brood).

Cultural and physical controls

- Implement practices promoting strong and healthy colonies, such as regular colony inspections, adequate nutrition and a clean hive environment.
- Only keep strong colonies. If colonies become weak, check that they are free from diseases and merge them (see 'I have two weak colonies; how can I merge them?').
- Use non-chemical methods to control pests, such as screened bottom boards to manage *Varroa* mite populations or freezing frames to prevent wax moth larvae.

Biological controls

- Foster a diverse and healthy ecosystem around your apiary that supports natural enemies of pests, such as predatory mites or beetles that prey on wax moth larvae.

Mechanical and behavioural controls

- Use traps strategically to capture pests like wax moths or small hive beetles.
- Use queens that show resistance to pests and diseases.

Chemical controls (as a last resort)

- Minimize the use of pesticides that can harm beneficial insects or contaminate hive products.

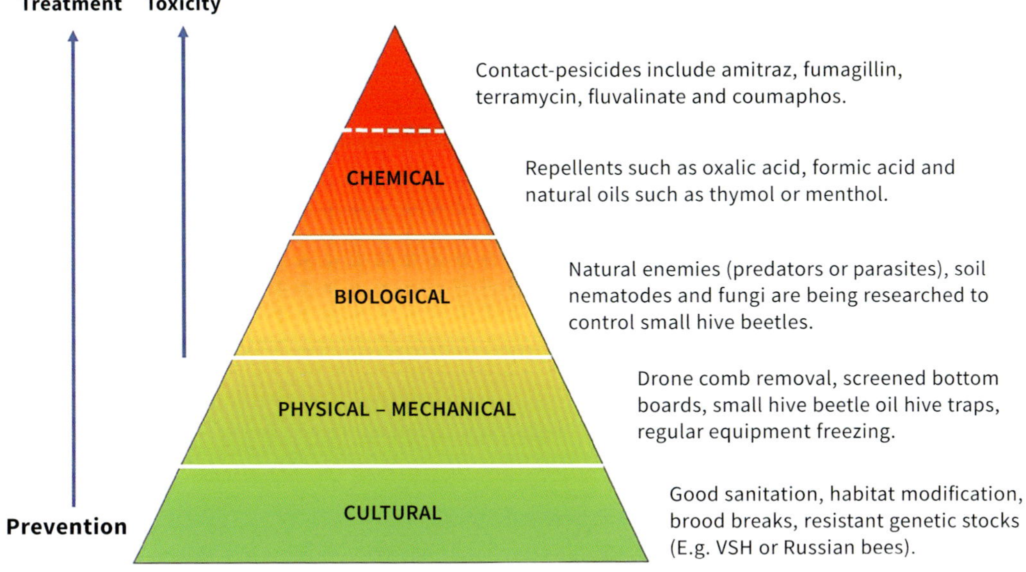

An integrated pest management strategy (IPM) needs to be designed and implemented for your apiary to minimize the use of chemicals and the resulting development of resistance

- If necessary, apply pesticides approved for beekeeping according to the label instructions.

Monitoring and evaluation

- Continuously monitor the effectiveness of your IPM strategy through hive inspections and pest monitoring.
- Modify your IPM strategy based on seasonal changes, new pest outbreaks or changes in bee colony health.

Record keeping and documentation

- Keep thorough records of all pest management activities, including dates of inspections, pest levels, treatments applied and the outcome of treatments. This is often a legal requirement.
- Use records to learn from your successes and failures in pest management.

Education and training

- **Keep up-to-date with the latest research and recommendations for bee health and pest management techniques.**
- **Educate others in your apiary community about IPM principles and practices.**

By following these steps, you can develop a comprehensive IPM strategy for your apiary that prioritizes the health of your honey bees while effectively managing pests and diseases. Remember, IPM is a proactive and holistic approach that aims to prevent problems before they occur rather than reacting after damage has already been done.

What should I do when my hive is infested with *Tropilaelaps?*

Problem: *Tropilaelaps* mites infest my hive. Urgent action is needed to manage this infestation.

Solution

Tropilaelaps mercedesae mites are a serious pest that can cause significant damage to honey bee colonies. They feed on the fat bodies of brood but not adult bees, leading to weakened bees, reduced brood production and sometimes colony collapse if the infestation is not addressed promptly.

In many ways, *Tropilaelaps* is similar to *Varroa*. *Tropilaelaps* is slightly smaller and has a less rounded body than *Varroa*. Like *Varroa*, it spends most of its life in capped pupal cells, its reproductive phase. *Tropilaelaps* mites only emerge when the adult pupa emerges (its dispersal phase), and the mite starts looking for an uncapped larval cell to enter to continue its life cycle. Unlike *Varroa*, *Tropilaelaps*, although it attaches itself to an adult bee when looking for an uncapped pupal cell, cannot use the adult bee for food since its mouthparts cannot penetrate the body of an adult

Left: *Tropilaelaps* on pupae with a deformed bee at the top left. *Tropolaelaps* mites also carry viruses that are lethal to bees.

Below: Female *Tropilaelaps mercedesae* and *Varroa destructor* on a pupa for size comparison.

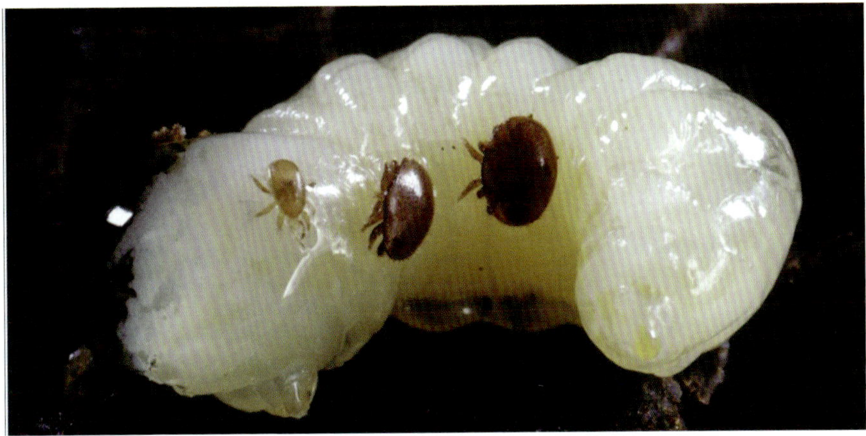

bee. Hence, adult *Tropilaelaps* cannot live more than three days outside a capped pupal cell since they cannot obtain food.

Tropilaelaps is considered a more significant problem for colony health than *Varroa* because it has a shorter reproduction cycle due to its reduced time in its dispersal stage, living phoretically on adult bees (three days vs five days for *Varroa*) and can thus overrun a colony with mature female mites quicker than *Varroa*. *Tropilaelaps* carries the same viruses as *Varroa*, including the deformed wing and black queen cell viruses, and poses the same risks to a colony as *Varroa*.

The steps needed to manage *Tropilaelaps* are mostly the same as those required to manage *Varroa*. Since *Tropilaelaps* has not been detected in Europe or North America, the same amount of research has not been done to manage this pest, and only limited recommendations can be made.

Caging the queen for up to 28 days, together with treatment with formic acid or, less effectively, oxalic acid, has been shown to significantly reduce the number of mites present in a colony. Formic acid will penetrate the capped surface of a pupal cell, where the mite spends most of its life. Effective treatment includes minimal capped brood; treatment with formic acid is thus best applied either during winter when the queen is laying fewer eggs or during queen caging when the queen is not laying.

Tropilaelaps mercedesae is spreading out of Asia and has already reached Georgia (country), Azerbaijan and southwest Russia (November 2024). The mite is spreading faster than anticipated around commercial colonies near the Black Sea, and we can expect to detect the mite in Europe in a few years. This would pose a severe financial challenge to many beekeeping operations in Europe.

Tropilaelaps may be challenging to identify due their physical similarity to *Varroa*, Pollen mites (*Mellitiphis alvearius*) and Braula fly (*Braula coeca*).

Dealing with *Tropilaelaps* mites can be challenging, but with careful management and early intervention, you can help your colony recover and reduce the impact of the infestation. Remember that managing a healthy, strong colony with good hygiene practices and timely interventions is your best defence against this and other bee pests.

See 'How can I manage *Varroa* by caging the queen?'

What is Colony Collapse Disorder, and how can I prevent it?

Problem: **I have read a lot about Colony Collapse Disorder (CCD) and wonder how to tell if my colony is affected and how to manage it.**

Solution

Unexplained colony losses have been reported since beekeeping records began. In the spring of 2007, several beekeepers in North America reported that in many of their colonies, all or most of the bees had left. Sometimes, the hive was empty; other times, only the queen and a few workers were present. Colony Collapse Disorder (CCD) is a name coined to describe these bee disappearances. Despite considerable efforts, the precise cause of CCD remains unclear. CCD is now a blanket term describing various conditions where many bees unexpectedly leave a hive. Colonies affected by CCD are sometimes infested with *Varroa destructor* and *Nosema ceranae*. Other possible causes have been speculated to be poor nutrition and pesticides. If you believe a colony is affected by CCD, recommendations for management include the following:

- **Do not merge a healthy colony with a collapsing colony.**
- **If you have a colony that has collapsed, destroy the hive in case it is infected with pathogens so that it cannot be used again.**
- **Contact your local government or the state apicultural officer for advice.**
- **Depending on the possible cause, treatment may be recommended. However, antibiotics can only be obtained with a prescription in many countries.**

What are the most common pathogens about which I should be concerned?

Problem: **I hear a lot about inspecting for pathogens and managing pests and diseases, but the literature lists many pathogens. I need to know which pests and diseases are the most common and which I should be concerned about.**

Solution

The primary pathogen affecting most colonies is the *Varroa* mite. Managing *Varroa* takes time and requires continuous monitoring of your colonies. *Varroa* is discussed in Chapter 18, '*Varroa* Mites'. A similar mite, *Tropilaelaps*, native to parts of Asia, would be equally serious if it broke quarantine and entered Europe, North America, Australia or New Zealand. In 2024, *Tropilalealps* was found in Azerbaijan, Southern Russia and Georgia (country), and will eventually spread to Europe.

The next most harmful pathogen is the bacterium that causes American foulbrood (AFB), *Paenibacillus larvae*. American foulbrood is discussed in 'What should I do if my colony shows signs of AFB?' Other pathogens are European foulbrood (EFB), Chalkbrood, *Nosema apis, Nosema cerana*; small hive beetles; and wax moths. Each of these is discussed later in this chapter.

What should I do if my colony shows signs of AFB?

Problem: I must identify and manage American foulbrood (AFB) in my colony.

Solution

American foulbrood (AFB), is a severe bacterial infection caused by the bacterium *Paenibacillus larvae*. European foulbrood is not as severe as AFB, although it still needs management. Both pathogens infect the brood. Infection with AFB usually shows signs after brood cells are capped. Infection with EFB usually shows signs before cells are capped, but this is not always the case. Since infection with AFB may be mistaken for other diseases, multiple methods should be used to confirm the diagnosis. These include:

Visual inspection:

- **Capped cells are perforated, sunken and greasy-looking, often containing a caramel-coloured liquid; late-stage larvae or early-stage pupae look sickly and discoloured (coffee brown to dark brown).**
- **A dark scale (desiccated larval remains) is on the lower cell wall. It is stuck tightly and is not removable without comb damage.**

- Extended pupal proboscis (false tongue) stretching from the lower to the upper cell wall — unique to AFB but rarely present.
- Distinctive foul odour — not always present and not readily identifiable to those not familiar with the odour.
- As the larvae continue to die, many cappings darken and develop holes. The holes are generally jagged and off-centre.

Conduct the 'ropiness'/ matchstick test.

- Select a brood cell that looks infected but not dehydrated (the prepupa/pupa structure is still evident and gooey).
- Use a matchstick or similar wooden implement to swirl and slowly withdraw the cell's contents. If the contents rope about 2 cm in length (>1 inch), the cell will most likely be infected with AFB. If the cell contents do not rope out but break and return to the cell, the infection is probably EFB.

What are the most common pathogens about which I should be concerned?.

- See 'How do I perform the Holst milk test to identify AFB?'

Lateral flow test kits.

- Similar to nasal tests that detect COVID, lateral flow test kits can be purchased online and from beekeeping supply stores. A sample of suspected infected brood is crushed and mixed with a reagent; the mixture is stirred, and the resulting mix is placed on a small test strip, indicating whether the brood is infected by the presence or absence of one or two stripes.
- Two separate test kits are available, one for AFB and EFB.

Laboratory identification.

- A sample consisting of a smear of a suspected infected capped brood cell on a light microscope slide can be sent to a diagnostic lab for identification.

- **A clear, liquid honey sample can be used by specialised laboratories to identify AFB. Many beekeepers systematically test a sample of the honey they collect at harvesting time. This gives them peace of mind that their apiary is safe from this disease.**

Treatment

The authors believe a colony with AFB should be destroyed, and the hive should be burnt and buried, or irradiated. This is a legal requirement in many countries, including Australia. Depending on the country in which you live or which state in the US, AFB may be treated with antibiotics available from beekeeping supply stores or by prescription from a veterinarian or government apiary inspector. Antibiotic treatment does not kill dormant AFB spores; and reinfection may occur many years later. Therefore, in areas where the use of antibiotics is allowed, it becomes very difficult to find any hives without at least some AFB spores, and abruptly stopping treatment often results in an outbreak of AFB. Under these conditions, burning of infected hives may not be practical, as the losses would be too high.

An alternative method frequently used in the UK, the US and parts of

1. Sunken, capped brood cells indicate infection with AFB.

2. Using the matchstick test, the residue from an infected larva can be strung out about 2 cm. The residue would not string out if the larva were infected with EFB. Sunken capping is often a sign of infection with AFB.

3. Healthy brood with no signs of disease.

Europe to manage AFB (and EFB) is the shook method, discussed in the section 'What is the shook swarm method to control EFB?' However, this method is illegal in countries such as Australia, where the colony must be destroyed.

AFB is a highly infectious pathogen that may remain dormant in a hive for many years. Equipment, including hive tools used during inspections, must be sterilized. Hive tools can be sterilized by soaking them in a strong disinfectant or placing them in a smouldering smoker.

How do I perform the Holst milk test to identify AFB?

Problem: I want to differentiate EFB from AFB in my hive.

Solution

The Milk test, also called the Holst test, is a simple and inexpensive test to determine whether a colony is infected with AFB but does not show a result if it is infected with EFB.

The Holst milk Test works best with active larvae infections but can also work on scales (i.e., the dark deposit left in cells after the bees have emerged). The proteolytic enzymes released by AFB bacteria break down many proteins, including milk proteins. The Holst milk test will be positive if a sample has enzymes that break down milk proteins — unique to AFB. In this test, suspect larvae are added to diluted milk. If AFB is present, the milk proteins will break down, and the solution will turn clear brown, losing its opaque 'milkiness.'

Materials needed:

- **Sample tubes.**
- **A small stick to collect larvae (you can use a matchstick from the matchstick test).**

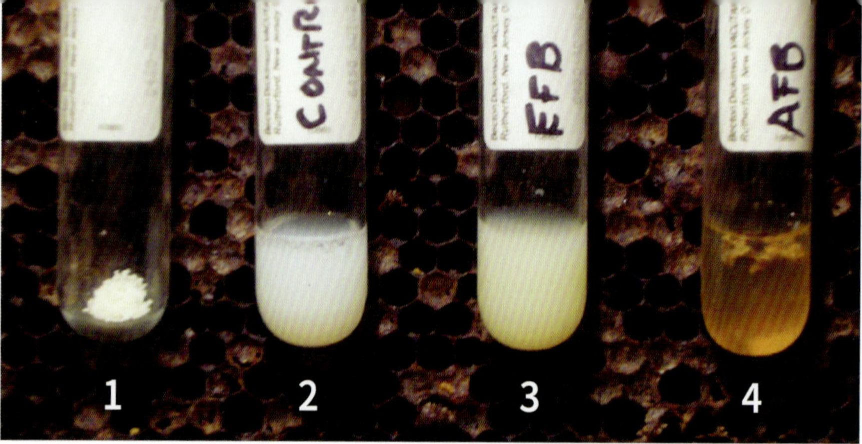

Results from a Holst milk test.

1. Tube 1 shows milk powder
2. Tube 2 shows milk powder dissolved in water or milk diluted with water.
3. Crushed larvae infected with EFB, stirred into the diluted milk solution, and left for 5 minutes, remain the same colour.
4. If the larvae are infected with AFB, the milk solution turns to a light brown, tea-like, colour after 5 minutes.

- Non-fat dry milk powder. Fresh milk may also be used and diluted with water. Kits can be made beforehand — place a small amount of non-fat dry milk and a stick at the bottom of the tube. The kits are fine to leave in your equipment until needed.

Steps to perform the test:

1. Prepare the diluted milk:

- Place a small amount of non-fat dry milk into the bottom of a tube. The actual amount does not need to be precise. You can also use fresh skim milk.
- Dilute the milk with water so it is still barely cloudy. If the amount of milk in the tube is a bit high, the results will be less clear but still evident. If you have fresh skimmed milk, you can dilute it in half and then half again. The test works best with milk at a dilution that is just cloudy.
- If you use dry milk, add water, cap the tube and shake to make a uniform liquid. Warmer water will allow the reaction to occur more quickly. If available, use warm water up to 65 °C (150°F), the temperature of hot coffee).

2. Collect the content of a cell to be tested:

- You can use the content of a capped larva, your matchstick test results or a dried scale.
- The more samples you add, the faster the reaction will occur. The enzyme is absent in young larvae, so older larvae or pupae should be used.
- Remember, you are looking for the enzyme responsible for breaking down larvae, so the grosser the dead larvae the better.

3. Let the enzymatic reaction occur and read the results:

- The reaction will not be immediate and will depend on the amount of sample, the dilution of the milk and the temperature. Put the tubes in a warm place, like your pocket, for approximately 20 minutes.
- Examine your results. The Holst milk test will confirm if the sample was positive for AFB. This test could be made more obvious using more diluted milk, warmer water and sample material. The positive sample will appear like iced tea when less milk is used.

What should I do if my colony shows signs of EFB?

Problem: My colony shows signs of European foulbrood (EFB), and I wonder what the best way to manage this infection is.

Solution

Since infection with EFB may be mistaken for other diseases, multiple methods should be used to confirm the diagnosis:

Visual inspection:

- Infection may be present in uncapped cells before the pupal stage.
- In capped cells, capping may be sunken with holes in the capping.

- Infected larvae may be yellowish, grey or brown.
- Larvae twisted in cell (corkscrew shape).
- Larvae may be transparent with their internal structure visible – the trachea may be seen.
- Larvae display a loss of internal pressure and appear soft in their structure.
- Dead larvae may form a dark rubbery scale on the bottom cell wall and be easily removed from the cell without damage.
- May be odourless or mildly sour. It may be mistaken for AFB for those not familiar with diagnosing the disease.

Lateral flow test kits

- Test kits called lateral flow devices, similar in shape to nasal tests that detect COVID, can be purchased online and from beekeeping supply stores. A sample of suspected infected brood is crushed and mixed with a reagent; the mixture is stirred, and the resulting mix is placed on a small test strip, indicating whether the brood is infected by the presence or absence of one or two stripes.
- Two test kits are available, one for AFB and the other for EFB, although these test kits are not interchangeable.

Differential diagnosis and co-infections

European foulbrood disease is often mistaken for other diseases, including AFB, Sacbrood and parasitic mite syndrome (PMS). It is often helpful to rule out other diseases:

- High *Varroa* mite counts — the colony may display symptoms of PMS.
- Use the rope test or Holst milk test to detect infection with AFB.
- The presence of chalk-like mummies indicates an infection with chalkbrood.

Be aware that co-infection of EFB with chalkbrood and Sacbrood is common.

Seasonality can aid in diagnosis.

- EFB is commonly detected following a strong honey flow (late spring in the UK and northern states in the US). Larval disease in late autumn is more likely related to high *Varroa* mite levels.

Laboratory identification.

- Send a sample to a diagnostic lab

Treatment

Colonies infected with EFB may spontaneously recover, particularly if the colony was strong. Periodically inspect all brood frames of an infected colony, noting the level of infection (approximate number of diseased cells/hive). If the colony shows signs of recovery, no further action may need to be taken. If the infection persists, the following alternatives may be used:

The bacteria that cause EFB thrive by digesting food in the larva's gut, starving the larva. Supplementary feeding with syrup and pollen substitutes may provide the larva with additional nutrition.

Use the shook swarm method to rehouse the colony into a new hive. See 'What is the shook swarm method for controlling EFB?' The shook-swarm method may be used with antibiotics, but this may not be necessary.

Antibiotics can be used to treat colonies that show signs of EFB and other hives in that yard, as the bacteria is

Top: Larvae showing signs of infection with EFB.

Right: Test kits are available to diagnose EFB. A different kit needs to be used to test for AFB.

generally present in adult bees in other nearby hives if the bacterium is present. This differs from a colony infected with AFB, where the authors recommend the colony and hive be destroyed. Prophylactic use of antibiotics for the prevention of EFB is illegal in some countries, such as Australia, while therapeutic use is allowed. Check with the relevant authorities in your area.

What is the shook swarm method to control EFB?

Problem: My colony is infected with European foulbrood, and I want to use the shook swarm method to manage the infection. I also wonder whether I can also use it for American foulbrood.

Solution

Studies have shown that the shook swarm method can effectively control EFB. Some countries also allow the shook swarm to be used to manage AFB; check with your beekeeping supply store or government apiary officer to see if it can be used to manage AFB. The technique may also control chalkbrood, *Nosema* and *Varroa*. This technique is not particularly suitable for smaller colonies. If shaking a small colony, shake it into a nucleus box.

The best time to carry out a shook swarm procedure is mid-spring, when plenty of floral resources are available, depending on the climate where you live. Before mid-spring, workers may have too little forage to draw new comb, and brood loss may be detrimental. If a shook swarm is carried out later in the summer, brood that would normally develop to collect food to take the colony through winter are destroyed. If the colony is strong, you can carry out this procedure earlier in the spring, but you will need to feed the bees syrup so they can build comb.

Shook swarm method:

1. Prepare a clean brood chamber filled with new frames of foundation, a clean floor, a crown board and a queen excluder.
2. Move the hive you intend to shake to one side and place the clean floor on the original hive-site. Put the queen excluder over the floor and then place the clean brood box containing foundation frames on top of the queen excluder.
3. Remove the centre four frames in the new hive and put them to one side. Flying bees will arrive at this new chamber.
4. Examine the old brood chamber and find the queen. Once the queen is found, place her between the frames of foundation in the new box or a queen cage so that she can be released into the new chamber when you have shaken the bees.
5. Shake all the bees from the old combs into the new chamber. This is done by holding a frame of bees about one-third of the way into the gap left between the foundations in the clean chamber. The frame is then moved quickly downward and suddenly stopped. Avoid jarring the comb against the chamber. The bees will fall off the comb, and any remaining on the comb can be brushed off.
6. Shake and brush bees on the old equipment into the new chamber.
7. When all the bees have been shaken into the new brood chamber, replace the frames of foundation that were removed.
8. Put the crown board in place.
9. Unless there is a strong nectar flow, feed with 'heavy' sugar syrup, i.e., 600 ml of water to 1 kg of white granulated sugar. It may be beneficial to delay feeding for two days. In this way, any contaminated nectar carried by the bees is used in comb building.
10. After about a week, when brood is present, remove the queen excluder and sterilize it appropriately.
11. Maintain feeding until all combs are drawn out. Check carefully at this point, as end combs may need to be turned around or moved one frame into the chamber. This is because bees find it hard to cluster and create wax on frames adjacent to the side of the brood box. If

EFB has been confirmed, old combs and frames must be destroyed.

See sections "What should I do if my colony shows signs of EFB?' and "What should I do if my colony shows signs of AFB?'

Shook method to manage EFB

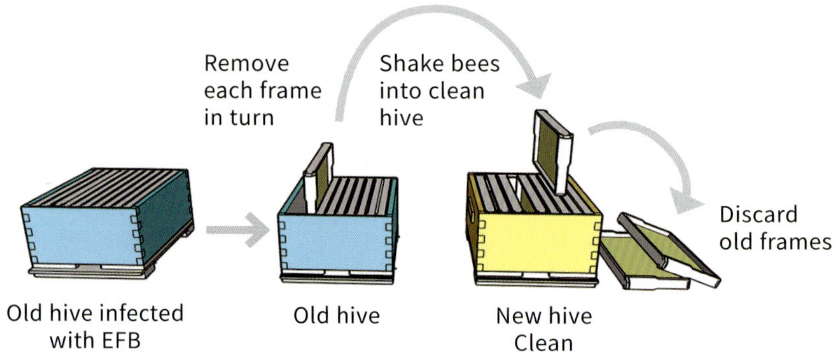

The shook swarm method to manage EFB. The process is the same for a top bar hive.

1. A hive body and colony are infected with EFB.
2. A clean hive is filled either with frames with new foundation or frames with a starter strip.
3. Frames from the old, infected hive are removed individually.
4. Bees on the frame are brushed into the new, clean hive.
5. Old, infected frames are discarded and destroyed.
6. The old, infected hive needs to be sterilized before using again.

What should I do when my hive is infested with small hive beetles?

Problem: There are small hive beetles in my hive, and I need to manage the infestation.

Solution

Small Hive Beetle (SHB), *Aethina tumida*, mainly affects weaker colonies; beekeepers can suppress the beetles' reproductive potential by maintaining strong bee colonies and keeping adult beetle populations low. Adult beetles live in areas of the hive box inaccessible to bees, such as under the hive mat or in places less often frequented by bees, like the corners of lids and cracks in the box's wood. In a strong hive, worker bees remove or drive the beetles from the hive. SHB tends to be more of a problem in hot humid climates such as in the tropics or subtropical areas. In temperate or Mediterranean climates, SHB does not generally cause hive losses. In many parts of the world it is not unusual to spot the occasional SHB running away when the hive mat is lifted. Strong hives generally keep small infestations in check and do not threaten the colony's health.

The larvae of SHB are often mistaken for wax moth larvae although there are significant differences between the two. (See section 'What should I do when comb is infested with wax moth' for a drawing that can differentiate between SHB larvae and Wax Moth larvae).

In some countries, chemical controls, including Checkmite+, are available for *Varroa* management but are limited in use against SHB. Good IPM practices in the apiary and extracting area are usually sufficient to minimize hive beetle infestations. Maintain a clean apiary and extracting area to reduce beetle attraction. A well-thought-out IPM plan will usually keep SHB infestations to a minimum.

Although supplementary feeding pollen during food shortages is a necessary part of good beekeeping, grease patties for tracheal mite control or adding protein supplement patties for spring build-up may increase SHB infestations since both adult beetles and larvae can feed on them.

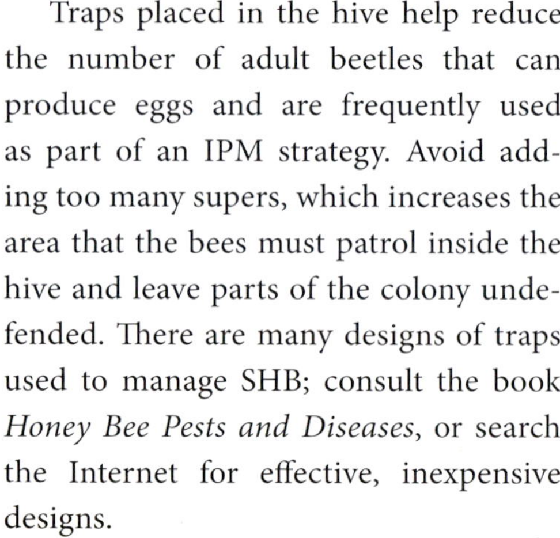

Traps placed in the hive help reduce the number of adult beetles that can produce eggs and are frequently used as part of an IPM strategy. Avoid adding too many supers, which increases the area that the bees must patrol inside the hive and leave parts of the colony undefended. There are many designs of traps used to manage SHB; consult the book *Honey Bee Pests and Diseases*, or search the Internet for effective, inexpensive designs.

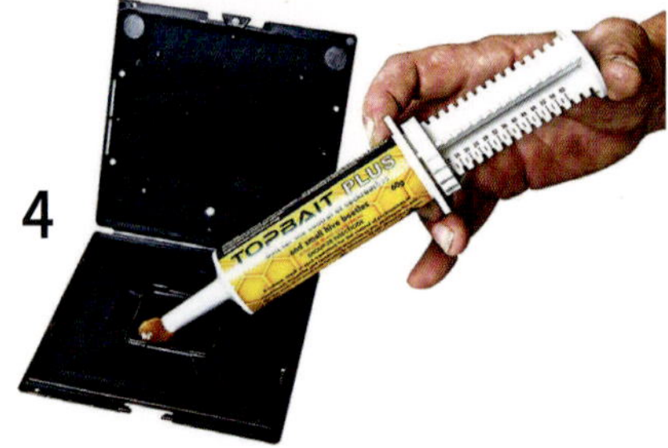

1. Small Hive Beetle larvae infesting a colony. This is called 'sliming'.

2. Adult SHB.

3. Traps are used to manage an infestation.

4. In some countries, chemicals are used inside a trap that bees cannot enter. Check with your local apiary inspector if this is allowed in your country.

What should I do when comb is infested with wax moth?

Problem: I want to prevent wax moths from destroying my frames when they are present in weak colonies or stored over the winter months.

Solution

The two wax moth species affecting honey bee comb are the Greater wax moth (*Galleria mellonella*) and the Lesser wax moth (*Achroia grisella*). A moth infestation can destroy a weak colony or stored frames very quickly. Preventive measures must occur before an infestation if the comb, and often the colony, is not to be destroyed. A strong colony can usually keep infestations under control. Ensure you have a strong and healthy colony by merging colonies if necessary (see 'I have two weak colonies; how can I merge them?'). If there is too much space for the number of bees, reduce the size of the hive by removing a super or brood box. These actions should keep moths out of the hive. Storing frames over winter allows the beekeeper to manage any infestation. Freeze each frame in the freezer for two days, then wrap each frame in a well-sealed plastic bag. This will kill any larvae and stop moths from accessing stored frames.

Chemicals that repel the moth are available for stored frames. Based on para dichlorobenzene, Para-Moth kills larvae and moths inside stored hives and in containers of stored frames. Check if this substance can be used in your country. Para dichlorobenzene cannot be used in supers containing honey or with frames that will be used to store honey.

Greater Wax Moth larvae are larger than SHB larvae, as shown on picture 4 on p. 296. Wax moth larvae are squishy when crushed, while SHB larvae are tougher. Wax moth larvae have three pairs of distinct front legs and a series of smaller proto-legs along the rest of their body; SHB larvae only have three sets of legs at the front.

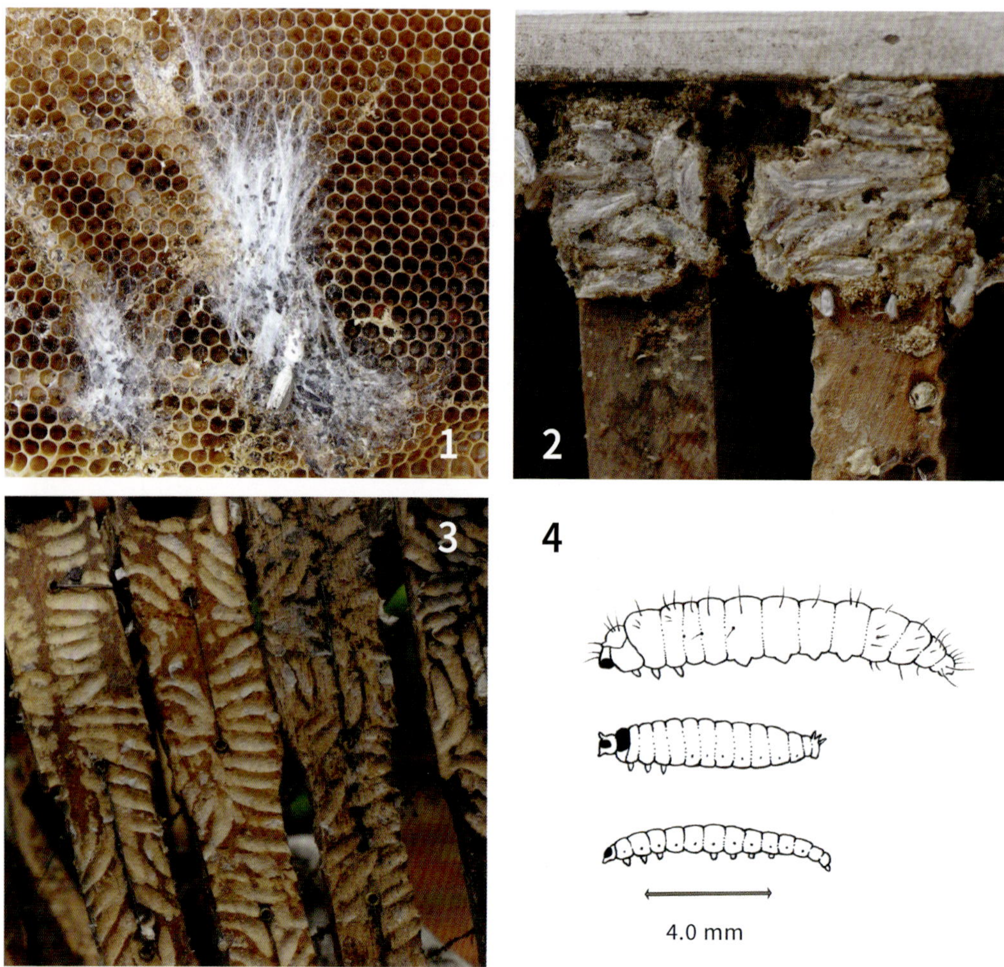

1. Comb damaged by wax moth.
2. Wax moth cocoons in a hive.
3. Moths can damage wooden hive bodies and frames.
4. Relative size of Greater Wax Moth larva (top), SHB larva (center), and Lesser Wax Moth (bottom).

What should I do if my hive is infected with *Nosema*?

Problem: I believe my colony is infected with *Nosema*. I want to know how to test for and treat the disease.

Solution

Nosema is a spore-forming fungus, not a bacterium or virus. There are two forms: *Nosema apis* and *Nosema ceranae*. Infection with *Nosema* is frequently called nosemosis or nosema disease. *N. apis* is believed to have originated in European honey bees (*Apis mellifera*), while *N. ceranae* is believed to have evolved more recently as an infection of Asian honey bees (*Apis cerana*). *N. ceranae* appears more harmful than *N. apis*. Nosemosis is often called a 'silent killer' because there are frequently no visible signs, although honey production may be lower than expected.

Infection of adult bees at a young age can cause them to have difficulty digesting food for the rest of their lives. Infected bees often do not produce brood food/royal jelly and may skip the brood-rearing stage, becoming forager bees at a young age. Infected bees frequently have a shortened adult lifespan. When queen bees become infected, they also have reduced lifespans and cease to lay eggs. These impacts cause reduced

Although many beekeepers believe that diarrhea on the front of a hive is a sign of infection with *Nosema apis*, this is not often the case, and if present, there may be other causes.

colony health, population and performance, which can ultimately result in the colony dying.

Nosema is best managed by maintaining strong, healthy colonies. To minimize infection, ensure that colonies have a good food supply and that a young queen heads the colony. Replacing brood comb every three to four years also helps keep the colony free of pathogens. Infection with *Nosema* frequently occurs during winter. When getting hives ready for winter, always ensure that there are no excess boxes on the hive and sufficient honey and pollen to last the colony over winter. If not, feed pollen supplements and syrup until there is an adequate level of food stored. Hives should be placed in a sunny position, free from wind, over late autumn, winter and early spring to help keep the colony warm.

No field test is available, so beekeepers need to monitor their colonies for indicators such as dysentery, reduced brood production, reduced honey production or population declines, particularly if there is no other reason for these symptoms.

What should I do when my hive is infected with chalkbrood?

Problem: **My hive is infected with chalkbrood, and I can see white and black mummies outside the hive and amongst the brood. I wonder how I should manage this disease.**

Solution

Infection with chalkbrood, caused by the fungus *Ascosphaera apis*, is easily detected since the mummies of dead larvae, coloured white or black, are found outside the hive entrance. Chalkbrood is most common in colonies already stressed by factors such as poor weather, low colony population, lack of food or mite infestation. There are no chemical treatments available for chalkbrood management.

1. Dead bee mummies infected with chalkbrood. Note the colour may either be white or dark-grey/blackish.

2. Chalkbrood is a fungus that breaks through the larva's skin. This photo shows the chalkbrood mycelium (white threads) occupying the entire cell volume outside the larval mummy.

3. Mummies at the front of the hive are usually the first sign that a colony is infected with chalkbrood.

Treatment involves managing the colony's other stress factors, such as controlling other pathogens, regularly replacing older brood comb and cleaning all mummies off the bottom board. Chalkbrood spores remain on the combs for many years, so you must remove and destroy heavily infected combs. The workers' cleaning behaviour is critical in preventing chalkbrood. If the workers quickly remove mummies from the hive, they can contain the infection. If you find combs with many mummies, the bees are not cleaning the hive effectively, and you need to requeen with a resistant queen that displays good hygienic behaviour. Hygienic behaviour is a characteristic that queen-rearers select for.

What should I do when my hive is infested with tracheal mites?

Problem: My colony is infested with tracheal mites; how should I manage this infestation?

Solution

The tracheal mite (*Acarapis woodi*) is a microscopic, internal mite that spends most of its life within adult bees' tracheae or breathing tubes. Infestations may result in weakened and sick colonies that do not work as hard, resulting in reduced nectar and pollen collection and significantly reduced lifespan. A colony can die if infested with tracheal mites and is also stressed by other problems, such as lack of food or *Varroa*. Once a honey bee colony is infested with tracheal mites, it remains infested, with impacts more significant over winter and early spring, contributing to high colony losses in severe cases.

Since the introduction of *Varroa*, the same hard and soft chemicals used to manage *Varroa* appear to have kept tracheal mite infestation under control. In some regions where queen trapping is used to manage *Varroa*, with little or no use of chemicals, tracheal mites may become a problem, requiring, unfortunately, the use of thymol.

Grease patties used to manage tracheal mites placed on top of frames in the brood box.

In countries with tracheal mites, integrated pest management (IPM) can help control infestations. This includes re-queening colonies susceptible to the disease, as colonies with young, active laying queens appear to cope better with mite infestations than older queens. Tracheal mite-resistant honey bee strains (e.g., Buckfast or Primorsky bees) can also be used. Tracheal mites are believed to have evolved with honey bees (*Apis mellifera*), and because of this, some strains of bees appear to be highly resistant to the mite.

In the UK, there is not an approved treatment for tracheal mites. In the US however, grease patties are frequently used to manage infestations. They are made from one-part solid vegetable oil and three parts sugar. While eating the sugar, bees get coated in oil, which protects them from mite infestation. Mated female mites have difficulty attaching to the outside of adult bees and transferring to other bees. Greased patties, if allowed in your country, have been shown to work best in early spring and autumn. Organic chemical treatments containing thymol gel or formic acid are also highly effective in controlling mites. Management using a combination of thymol and grease patties is usually sufficient to control an infestation.

What are 'Africanized bees', and how do I prevent them in my hive?

Problem: I live in a region (South America, Central America and Southern USA) where Africanized bees, also known as 'killer bees', are present. My bees are very aggressive, and I suspect Africanized bees have taken over my hive. I would like to know what to do.

Solution

Africanized bees originate from Brazil following the accidental release of a strain of bees created by researchers at the University of Sao Paulo by artificially inseminating an *Apis mellifera scutellata* queen, originating from Africa, with sperm from *A. m. ligustica*, originating from Italy. These aggressive bees have spread from tropical Brazil, north to the Southern states of the US. Since they are best adapted to warm climates, their northward spread was stopped by the cold winters typically found in the northern US.

Their relative resistance to *Varroa* mites, combined with their exceptional evolutionary fitness, partly because of their aggressive behaviour and propensity to swarm, often resulted in them taking over many feral hive sites in South and Central America and the Southern US. In these areas, local feral and managed bees mate with feral Africanized bees, resulting in all colonies possibly possessing some Africanized genetics. As a result, bees and queens bred locally are often restricted from sale.

While it is possible to formally identify Africanized bees based on genetic markers and vein patterns in the wings, it is much easier to identify them through their highly aggressive behaviour. Africanized bees' behaviour includes rapid deployment of multiple bees that sting intruders at the slightest provocation, bees that pursue intruders for a long distance from the hive, much lower sensitivity to smoking and a propensity for widespread stinging once an attack has started.

Extreme caution is required when dealing with Africanized bees, as the outcome of an attack on an unprepared person can be fatal. Special bee suits can be used, along with additional precautions such as duct-taping the fabric of any joining pieces of protective clothing. In addition, much larger smokers control the attack pheromones released by the bees.

How can I disinfect my hive and equipment?

Problem: **I suspect or know I have an infectious disease in my hive and want to disinfect my equipment.**

Solution

Disinfecting means removing most germs, while sterilization means removing all microorganisms. Most of the methods discussed here are disinfecting, with only some resulting in sterilization. In most cases, disinfecting is sufficient. Still, there are exceptions, most notably AFB, which requires sterilization due to the pathogen-producing spores that are highly resistant to most disinfecting methods. Disinfecting your beehive is crucial for preventing disease and pests that can harm your bees. Here are some methods to disinfect different parts of your beehive:

Cleaning with hot water and soap

Remove all debris, wax and propolis from the hive parts using a hive tool. Wash the hive parts thoroughly with hot water and a mild soap. A wire brush can help remove stubborn residues, including propolis. Rinse with clean water. Allow the parts to dry completely before reassembling the hive parts. This is an easy method of disinfecting your equipment and should be done regularly, including between opening different hives. If your hives are close together, consider carrying multiple hive tools so that one is being sterilized while you use a clean tool.

Hot wax dipping

Many places that manufacture or sell hives produce hot wax dip hive parts. The assembled box is immersed in hot beeswax for five minutes, although sometimes microcrystalline wax is used. When the box is removed from the wax, it is usually painted immediately so that the cooling wax and paint are absorbed into the wood pores. A wax-dipped box can be left unpainted since the wax coating protects against weather and rotting wood. The wax coating kills many microorganisms due to its temperature during application. Also, it seals the wood so that microorganisms in the wood cannot escape into the colony and infect bees. This comes very close to sterilizing the equipment and is effective against AFB, for example.

Heat treating

Many beekeepers disinfect their hives by burning the inside using a butane or propane torch. This method is pretty effective. Still, in some parts of the world (e.g., Australia), it is not recognized as an acceptable way to treat AFB.

Sunlight

Leave the hive parts in direct sunlight for several hours or days. The UV rays can help kill pathogens. This is partially effective at disinfecting but is not a recognized way to treat AFB.

Chemical disinfectants

Mix 1 part household bleach with 10 parts water. Soak the hive components in the solution or wipe them down. Rinse thoroughly with clean water and allow the parts to dry completely, preferably exposed to sunlight.

Use a 5 per cent acetic acid (vinegar) solution. Wipe down the surfaces, then rinse and dry, preferably exposed to sunlight. This is a good way to disinfect a hive or other equipment, but it does not always lead to sterility.

Professional methods

Some beekeepers use gamma radiation to sterilize equipment, especially in more extensive operations or when dealing with severe disease outbreaks. It is acceptable for treating AFB. However, plastic parts, including zippers of bee suits, bee brushes and plastic queen excluders, are affected by the radiation and often become brittle.

Additional tips

Before reassembling, the hive parts must be dry and free from chemical residues. This can be done by leaving them in the sun for a few days. This will help ensure the health and safety of your bee colony.

CHAPTER 20

Other Problems

Asian hornet, *Vespa velutina.*

What should I do when I have chilled brood?

Problem: Opening the hive during cold conditions, and especially shaking bees off comb during such periods, can result in chilled brood and should be avoided.

Solution

Honey bees are remarkable because they can regulate the hive's temperature. They collect water from nearby and evaporate it to keep cool in hot weather. In cold weather, they use nectar and honey to fuel the movement of their wing muscles (disconnected from the wings) to produce heat. This keeps the temperature of the brood cells at 37 °C (100 °F). However, when exposed to the elements on cold days (below 18 °C or 65 °F), they can have difficulty maintaining and restoring that temperature once the beekeeper closes the hive again. This is particularly true when the weather is windy or wet.

Brood that has been cooled for some time can die off, often in an irregular pattern, or there might be a resurgence of chalkbrood. Chalkbrood is a fungal disease characterized by the appearance of white or dark grey solid dried pupae remains in the brood cells. Depending on the strain of bees (i.e., some with a high degree of hygiene), the workers will remove these dead, dried pupae and drop them in front of the hive. Hence, dried white or dark grey pupae in front of the hive may indicate a chilled brood. These dried, dark larvae may be confused with a Chalkbrood infection.

Prevention is easy: The beekeeper must resist the urge to inspect the hive during cold or wet weather. If an inspection is needed, it should be performed quickly without shaking the bees off the comb and in the late morning on a sunny/wind-still day (giving the bees plenty of time to heat the hive before dark and colder temperatures at night).

Keeping the hive as small as possible (i.e., using as few hive boxes as possible for the number of bees) can also help prevent brood chilling. Also, ensure that there are always ample supplies of honey or nectar in the hive during colder months.

If more than a few white/dark grey dried pupae are found in front of the hive, the best course of action is to quickly increase the number of bees by merging with another colony and replacing the queen, particularly one with a high degree of hygiene (ask the queen supplier). Check a few weeks later for the absence of dried white or dark grey pupae.

What is pesticide poisoning, and what can I do to prevent it?

Problem: I observe a larger-than-usual number of dead bees at the hive's entrance and wonder what it means and what to do about it.

Solution

Many dead bees outside a hive may be drones ejected by workers as winter approaches and food becomes scarce. First, check that the dead bees are drones; if not, suspect pesticide poisoning.

Poisoning bees with pesticides can be relatively regular, depending on where you keep your bees. In urban areas, it is not unusual for gardeners to use pesticides on vegetable gardens or flower beds. In rural areas, many farmers use pesticides at critical times of the year. Whichever the situation, considering prevailing winds, which drive pesticide drift and the flight distance of bees (typically up to 3 or 4 kilometres or about 1.5 miles). Pesticides used in any area up to five kilometres from the hive could be cause for concern. Generally, gardeners and farmers are pretty receptive to the idea that bees are affected by their pesticide usage. Because they are helpful to their crops, they may agree to let nearby beekeepers know when they will spray. A beekeeper informed of spraying activities in his/her area would be wise to close the hive entrance using a mesh (for ventilation) after providing some water (especially if the weather is predicted to be hot). Bees confined to the hive will be much less exposed to sprayed pesticides and, therefore, are more likely to survive. Alternatively, the beekeepers may move their hives safely from the spraying (greater than 5 kilometres). Because this is likely to be more than 3 kilometres away from the current position of the hive this may be a relatively simple endeavour (See ‘How can I move my hive to a different location?’).

How do I prevent mice from infesting my hive?

Problem: **Over winter, mice are getting into my hives and making nests, and I want to prevent this from happening.**

Solution

Mice have an uncanny ability to make themselves thin and, with very flexible joints, to squeeze through tiny gaps. To minimize the ability of mice to enter your hive, first ensure that all gaps, except for the main entrance, are sealed. Next, purchase a mouse grill and attach it over the main entrance. A mouse grill has small gaps that allow bees to pass but are too small to enable mice to pass. This is mainly a problem during cold winters when the bees are clustering and not protecting the entrance as effectively as during other times.

A mouse guard is an effective way to stop mice entering your hive during cold weather or in search of food.

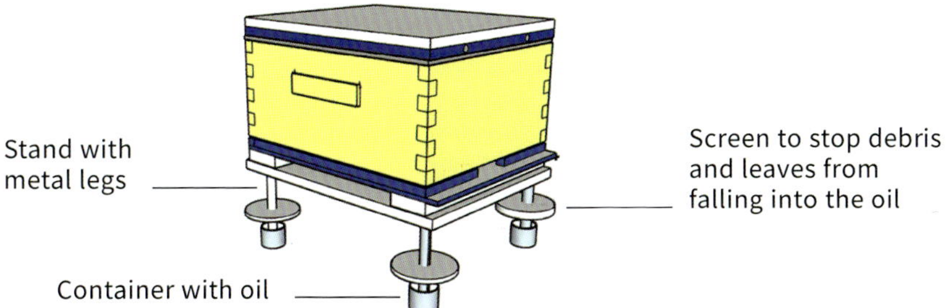

A stand with its legs placed in jars of oil to stop ants entering a hive. A lid is placed above the oil to keep debris and leaves out of the oil.

How do I prevent ants from infesting my hive?

Problem: I live in an area with many ants and want to prevent them from entering my hives.

Solution

Ants can be a problem in tropical and subtropical regions where they can cause the colony to abscond if things get out of hand. In temperate climates, ants are usually only irritating and do not bother the colony. Effective solutions include the following:

Scatter cinnamon powder around the hive and, if necessary, place it in the lid. Ants hate the smell of cinnamon, while bees tolerate it. Other natural substances that repel ants are peppermint and tea tree oil. Spray around the base of the hive and reapply if it rains.

The most effective solution is to raise the hive above the ground and apply a grease or liquid barrier so the ants cannot pass. If the hive is placed on a single large stump or post, apply grease around the post. If the hive is placed on a stand, place each of the stand's legs in a pail of water or oil. In either case, ants will not get past the liquid to reach the hive.

For minor infestations, particularly in a domestic situation, try placing ant traps containing borax near the hive, ensuring other domestic animals cannot access the bait.

How do I prevent my hives from being disturbed by large animals?

Problem: **Large animals such as bears and horses disturb my hives, and I want to protect my apiary.**

Solution

The best solution is to place an electric fence around the hives. A standard fence may not deter a determined attacker like a bear, but it will deter horses and most farm animals. Large animals tend to rub against the hive, resulting in aggressive bees that can sting the animal and/or push the hive over, resulting in being stung by the bees.

A fence around an apiary will deter most large animals from entering.

Can I keep bees together with other livestock or pets?

Problem: I want to keep bees in close contact with dogs, chickens and other domestic or farm animals.

Solution

You can keep bees near dogs, chickens and other domestic or farm animals, but there are some considerations to keep in mind to ensure the safety and well-being of both the animals and the bees:

- Ensure the beehives are placed where animals cannot easily disturb them. Bees generally prefer quiet areas away from constant disturbances.
- Bees need access to water, so providing a separate water source away from where animals drink can help prevent conflicts. Some animals, however, can be placed near each other. Hives are often located inside a chicken coop since they get on well and do not disturb each other.
- Consider fencing around the beehives to deter animals from investigating too closely. This can help prevent accidental disturbances or confrontations.
- Know your animals' behaviours. Some dogs, for example, may be curious about the bees and inadvertently disturb the hives if not adequately trained or supervised.
- Regularly monitor the animals and the bees for any signs of stress or disturbance. This allows you to intervene if necessary to prevent issues.
- Be aware of your bees' flight path. Ensure they have clear access to forage without flying directly over areas where animals are likely to be disturbed.

- **Proper hive management, including regular inspections and maintenance, can help prevent swarming or other behaviours that might alarm nearby animals.**

By taking these precautions and being attentive to the interactions between your bees and other animals, you can create an environment where both can coexist peacefully and safely.

What should I do if wasps or hornets attack my colony?

Problem: Hornets live near where I keep my hives, and I want to prevent them from attacking and robbing my colonies.

Solution

Invasive hornets have become a problem for beekeepers in Europe, the UK and North America, weakening and killing colonies. As these insects increase in population size and spread further, they will become an even more significant threat to honey bees. Bees are already threatened by recent invasive pathogens such as *Varroa*, Small Hive Beetle, *Nosema cerana*, tracheal mites and Deformed Wing Virus. In large numbers, managing the different species of introduced hornets adds work to beekeepers and makes the viability of the bee population more tenuous. Studies in France have shown that about half of the diet of a hornet nest consists of bees, a significant amount considering the low but increasing hornet population density.

There are many species of wasps and hornets, such as the invasive Asian or *yellow legged hornet (Vespa velutina)* in UK and Europe, or the Asian Giant Hornet (*Vespa mandarinia*) in North America, and whether they are likely to represent a problem for beekeepers depends on the geographical area in which the bees are kept. Seek advice from a local bee club or supplier of bee equipment. Wasps and hornets typically can kill bees at the

1. Making a hornet and wasp trap from a drink bottle.

2. Attaching trap to tree.

3. Using a trap to catch wasps, yellow jackets and hornets.

4. A Japanese method of restricting access of hornets to the colony. The height of the entrance is 6 mm (1/4 inch). Hornets and mice cannot pass this narrow entrance.

entrance and feed on honey and brood if they can access the inside of the hive. As a result, they may further weaken an already weak hive, and in the worst case, they will completely decimate the weak hive. Wasps will harvest adult bees, remove the head, legs and abdomen, and return to their nest with the thorax. Thus, a clear sign of an attack by wasps is the observation of bee heads at the hive's entrance.

Hornets typically stay near the colony's entrance, attacking foragers entering or leaving the hive or entering the hive themselves and feeding on larvae and adult bees. Bees can sense that hornets are outside and do not leave the colony to forage; hence, they risk starving even if a hornet does not kill them.

A practical solution in Europe is to attach a grill next to the hive's entrance. This grill allows bees to enter and leave but has a sufficiently small mesh size to prohibit hornets entering. The mesh is part of a larger box that fences off much of the front of the brood box . Bees detect the presence of a hornet and will enter and leave the cage on a different side from the hornet (see 'What should I do if my hive is being robbed by neighbouring hives?').

Another popular solution is to set a trap near the hive that attracts hornets but leave bees indifferent. These traps are relatively easy to make or can be purchased quite cheaply. Lures used in these traps include dried fish, beer and apple cider vinegar. Commercial traps, including VespaCatch® and Tap-trap®, purchased locally will have a lure recipe that works well for the local species. While wasps are mainly a problem in autumn when wasps accumulate food stores for winter, traps work best in spring when queens are out to form new nests. DIY traps can be made of water bottles containing meat grease (for protein), jam and vinegar, or other attractants. Water in the bottle should include a small amount of detergent to interfere with the hornet's ability to leave, causing them to drown. Care must be taken when designing the water bottle trap so that it does not also attract bees.

Established wasps and hornet nests can also be eliminated using insecticides. However, these insecticides may also kill your bees, even in small amounts that may be blown into the environment. It is, therefore, necessary to close your hive when doing this, or even better, relocate the hive far from the wasp/hornet nest to be treated (see 'How can I move my hive to a different location?'). Also, wasps and hornets being exterminated can be very aggressive, and great care must be taken. Alternatively, employ a professional to do this job safely.

See 'What should I do if my hive is being robbed by neighbouring hives?'

ACKNOWLEDGMENTS

This book could not have been written without the assistance of many people. Beekeeping is wide, and no person can be an expert in all aspects of this fascinating activity. Many people generously donated significant time from their busy schedules to read and make suggestions to improve the text. In alphabetical order, I would like to thank:

Ben Moore, www.bensbees.com.au, for providing many images and checking large parts of the text.

Bron Woods, Bobs Beekeeping Supplies, Watsons Creek, Victoria, Australia, who read parts of the manuscript and provided many photographs.

Jane Murray, who generously read the manuscript twice, not once, and made many helpful suggestions, both about the content as well as grammar.

Kris Fricke, the editor of *Australasian Beekeeper*, read the entire manuscript and made many valuable suggestions about beekeeping to make the book more suitable for a North American reader.

Phil Breed is an experienced US beekeeper who owns Birdsong Bees in Nebraska, and works as a beekeeping technician at the University of Nebraska. Phil read the entire manuscript, made many valuable suggestions, and provided several photographs.

Tim Blumfield read the manuscript and made valuable suggestions that improved the final version.

PICTURE CREDITS

FRONT COVER: 1.4 Wyns, Daniel, 5.28 Hives in Field. Moore, Ben, 5.28 Hives in Field. Moore, Ben - Duplicate, 10.53c Wikipedia, 10.53d Wikipedia, 14.83 Moore, Ben, 17.87 Freepick, 17.88 Everrett, Priscilla / Luberto, Elenora, of Jemman Photography, Dreamstime, **PAGES**: xi Shutterstock, xii Prostockstudio/Dreamstime.com, 1 Chapter 1: Getting Started – Photo: Moore, Ben, 3 Wikipedia, 7 Wikipedia, 8 Dadant, 12 Wyns, Daniel, 14 Chapter 2: Equipment – Photo: Breed, Phil, 15a Anonymous, 15b Dunn, Fred, 16 Anonymous, 19 Owen, Robert, 21a Wikipedia, 21b Anonymous, 21c Anita - I may be able to find Anita´s last name when I return home on the 14th, 24 Anel, 27 Pierco, 30 Owen, Robert, 32.1 Owen, Robert, 32.2 Owen, Robert, 32.3 Owen, Robert, 32.4 Owen, Robert, 33.1 Anonymous, 33.2 Scheerlinck, JP, 32.3 Owen, Robert, 33.1 Anonymous, 33.2 Scheerlinck, JP, 35 Owen, Robert, 37 Owen, Robert, 39 Owen, Robert, 41.1 Owen, Robert, 41.2 Owen, Robert, 41.3 Owen, Robert, 42 Chapter 3: Looking After my Equipment – van Oort, Lucas, 43 Moore, Bens, 44.1 New Brunswick Botanic Gardens, 44.2 Breed, Phil, 46 Bush, Michael, 48 Whirrakee Beekeeping / BetterBee, 51 Owen, Robert, 52 Chapter 4: Clothing – Photo: Getty Images, 53 Stockphotos, 54 Owen, Robert, 55 Breed, Phil, 56 Owen, Robert, 58 Owen, Robert, 59 Chapter 5: Locating your Hives – Photo: Owen, Robert, 60 Moore, Ben, 63.1 Barry Cooper, 63.2 Owen, Robert, 63.3 Farrell, Cormac, 65.1 Owen, Robert, 65.2 Scheerlinck, JP, 65.3 Owen, Robert, 66 Chapter 6: First Hive Opening – Bouffard, Maxime, 67 Owen, Robert, 68 Anita – I may be able to find Anita´s last name when I return home on the 14th, 70 Owen, Robert, 72 Owen, Robert, 73 Chapter 7: Spring & Summer Management – Shutterstock, 75 Owen, Robert, 77 Owen, Robert, 82 Owen, Robert, 89 Chapter 8: Autumn & Winter Management – Breed, Phil, 89 Zawislak, Jon, 92.1 keepingbackyardbees, 92.2 Walmart, 94 NOD Apiaries, 96 Freeze Frame, 97 Chapter 9: Feeding Bees – Mora, Alexiz, 99 Agrifutures, 103.1 Scheerlinck, JP, 103.2 Scheerlinck, JP, 103.3 Scheerlinck, JP, 103.4 Scheerlinck, JP, 106.1 Owen, Robert, 106.2 Owen, Robert, 106.3 Pierco, 106.4 Owen, Robert, 106.5 Owen, Robert, 108 Super Bee, 110 Chapter 10: My Bees – Breed, Phil, 111 Moore, Ben, 112 Glenn Apiaries, 114 Carolina Honey Bee, 118 Anderson, Dennis, 119.1 Owen, Robert, 119.2 Bentley, Michael, 121 Watling, Sue, 123 Snyder, Robert, 125.1 Wikipedia, 125.2 Wikipedia, 125.3 Wikipedia, 125.4 Wikipedia, 132 Owen, Robert, 134 Chapter 11: Swarming - Breed, Phil, 135 Milkwood, 137 Owen, Robert, 140 Owen, Robert, 144.1 Milkwood, 144.2 Milkwood, 144.3 Milkwood, 144.4 Breed, Phil, 145 Happy Hollow Honey, 147.1 Moore, Ben, 147.2 Bee Equipment Australia, 148 Chapter 12: Queens – Moore, Ben, 149 Bentley, Michael, 153 University of Georgia, 156.1 Horner, Joe, 156.2 Horner, Joe, 156.3 Horner, Joe, 159.1 Everrett, Priscilla / Luberto, Elenora, of Jemman Photography, 159.2 Owen, Robert, 159.3 Scheerlinck, JP, 159.4 Owen, Robert, 165.1 Owen, Robert, 165.2 Owen, Robert, 166 Owen, Robert, 167 Chapter 13: Queen Rearing – Anonymous, 168 Owen, Robert, 169.1 Owen, Robert, 169.2 Owen, Robert, 169.3 Owen, Robert, 169.4 Owen, Robert, 171 Owen, Robert, 174.1 Owen, Robert, 177.1 Owen, Robert, 177.2 Owen, Robert, 177.3 Owen, Robert, 177.4 Owen, Robert, 180 Golden Bee Manitoba, 181.1 Anonymous, 181.2 Glenn Apiaries, 181.3 Owen, Robert, 182.1 Owen, Robert, 182.2 Owen, Robert, 185

Chapter 14: Harvesting Honey – Moore, Ben, 186 Breed, Phil, 187 Owen, Robert, 188.1 Scheerlinck, JP, 188.2 Scheerlinck, JP, 190.1 Owen, Robert, 190.2 Owen, Robert, 190.3 Owen, Robert, 190.4 Owen, Robert, 197 Moore, Ben, 200 Anonymous, Chapter 15: Natural Beekeeping – Bentley, Michael, 205 Voigt, Ronnie, 207.1 Voigt, Ronnie, 207.2 Farrell, Cormac, 207.3 Voigt, Ronnie, 212 David Dawson, 218.1 Scheerlinck, JP, 218.2 Owen, Robert, Chapter 16: Flow Hive - NSW DPI, 222 Dunn, Fred, 230 honeyflow.com, 231.1 Dunn, Fred, 231.2 Dunn, Fred, 231.3 Dunn, Fred, Chapter 17: Other Products – Sant´Anna, Roberta, 233 Freepick, 234 Everrett, Priscilla / Luberto, Elenora, of Jemman Photography, 235.1 Chariot Honeycomb, 235.2 Chariot Honeycomb, 238.1 Owen, Robert, 238.2 Owen, Robert, 239.1 Owen, Robert, 239.2 Owen, Robert, 239.3 Bentley, Michael, 241.1 Owen, Robert, 241.2 Owen, Robert, Chapter 18: Varroa – Moore, Ben, 241 CSIRO, 244.1 Louisiana State University, 244.2 Anonymous, 247 Owen, Robert, 248 Whirrakee Beekeeping, 250 Scientific Beekeeping, 252.1 Iowa State University, 252.2 Scientific Beekeeping, 256.1 Scheerlinck, JP, 256.2 Scheerlinck, JP, 256.3 Scheerlinck, JP, 256.4 Scheerlinck, JP, 259.1 Scheerlinck, JP, 259.2 Scheerlinck, JP, 259.3 Scheerlinck, JP, 259.4 Scheerlinck, JP, 261.1 FERA, 261.2 FERA, 263.1 Pierco, 263.2 Owen, Robert, 265 Owen, Robert, 270.1 Owen, Robert, 270.2 Lopez-Uribe Lab, 270.3 Lopez-Uribe Lab, 271 Veto Pharma, 272 Wikipedia, 273 University of Georgia, 274 Chapter 19: Other Pests – Moore, Ben, 275 USDA, 277 Owen, Robert, 279.1 CSIRO, 279.2 Anderson, Dennis, 284.1 USDA, 284.2 USDA, 284.3 Anonymous, 286 Scientific Beekeeping, 289.1 Anonymous, 289.2 Vita Phrama, 292 Owen, Robert, 294.1 Bugwood, 294.2 Bugwood, 294.3 Vita Pharma, 294.4 Vita Pharma / Top Bait, 296.1 Owen, Robert, 296.2 Owen, Robert, 296.3 Owen, Robert, 296.4 Owen, Robert, 297 Braberg, David, 299.1 USDA, 299.2 AFB.ORG, 299.3 Snyder, Robert, 300 Galena Farms, 305 Chapter 20: Other Problems - Wikipedia, 308 pxhere.com, 309 Owen, Robert, 310 Watling, Sue, 313.1 Owen, Robert, 313.2 Owen, Robert, 313.3 Anonymous, 313.4 Anonymous, 316/317 Dreamstime, 318 Aleksandr Zamuruev/Dreamstime.com, Endpapers: AndreyKuzmin/Dreamstime.com

BIBLIOGRAPHY

The Art of Beekeeping by Phil Chandler
The Australian Beekeeping Manual by Robert Owen
The Backyard Beekeeper by Kim Flottum
The Beekeeper's Bible by Richard A. Jones and Sharon Sweeney-Lynch
The Beekeeper's Lament: How One Man and Half a Billion Honey Bees Help Feed America by Hannah Nordhaus
The Beekeeper's Problem Solver by Deborah Horn
Beekeeping: A Practical Guide by Clive de Bruyn
Beekeeping for Dummies by Howland Blackiston
Bees: A Natural History by Laurence Packer
Bees and Beekeeping: A Year in the Life of an Apiary by Roger A. Morse and Nicole D. Morse
Beesource Beekeeping Basics by Beesource
The Complete Idiot's Guide to Beekeeping by Dean Stiglitz and Laurie Herboldsheimer
First Lessons in Beekeeping by C.P. Dadant
The Hive and the Honeybee by Brother Adam
Honeybee Democracy by Thomas D. Seeley
Honey Bee Pests and Diseases by Robert Owen, Jean-Pierre Y. Scheerlinck, Mark Stevenson
The Honey Bee: A Guide to Productive Beekeeping by B.L.P. Neve
Honeybee Health and Disease Management by Diana Sammataro and Alphonse A. Avitabile
Keeping Bees: A Complete Practical Guide by Stanley W. D. S.
Natural Beekeeping by Ross Conrad
The Practical Beekeeper: Beekeeping Naturally by Michael Bush
Queen of the Sun: What Are the Bees Telling Us? by Taggart Siegel and Jon Betz
The Secret Life of Bees by Sue Monk Kidd
The Wisdom of Bees: What the Hive Can Teach Business About Leadership, Efficiency, and Growth by Michael O'Malley

GLOSSARY

ABSCOND: This occurs when a colony of bees suddenly leaves a hive with few or no bees remaining behind. Absconding should not be confused with swarming; it is often caused by problems such as poor temperature/ventilation within the hive, mites, pests, lack of food or other intolerable problems.

ACUTE BEE PARALYSIS VIRUS (ABV): A viral infection of the honey bee that causes paralysis of adult bees.

AFRICANIZED HONEY BEE: A term used to describe the South African honey bee *Apis mellifera scutellata* or its hybrids, now found in North and South America. Africanized honey bees are known for their volatile nature. The term Africanized honey bee is used to differentiate the type of bee found in the Americas from African honey bees, which although of the same race are only found in Africa.

AFTER-SWARM: A swarm that leaves a colony with a virgin queen shortly after the first (or prime) swarm has departed. The first after-swarm is also called a secondary swarm while an after-swarm that leaves after the secondary swarm is also called a tertiary swarm.

AMERICAN FOUL BROOD (AFB): A lethal infection of honey bees with the bacterium *Paenibacillus larvae.*

ASIAN BEES: Of the eight currently identified species of honey bees, seven are only found in Asia, and these are called by the generic name Asian bees. The eighth type of bee, the European honey bee, *Apis mellifera,* was only found in Africa, Europe and parts of the Middle East before being transported globally by humans. It is, thus, not an Asian bee. Note that Asian bee is a generic term for seven species of honey bee found in Asia, while the term Asian Honey Bee, AHB, usually applies to the single species, *Apis cerana.*

ASIAN HONEY BEE: One of the eight species of honey bee found globally, the Asian honey bee (*Apis cerana*) is found across large areas of Asia and parts of Queensland.

BAIT HIVE: An empty hive left to attract swarms of bees. The term sometimes refers to hives left near ports or airports to attract swarms that have arrived by ship or plane and have eluded capture by quarantine. A bait hive may use the Nasonov pheromone to attract swarms.

BEE SPACE: A space between two frames that is big enough to allow two bees to pass each other while working on opposite frames. A bee space is about 9.5 mm wide, too small to encourage comb building, and too large to induce propolising activities.

BLACK BEES: *Apis mellifera mellifera,* sometimes called German bees.

BRACE COMB: Pieces of seemingly random comb that connect hive parts, such as between two frames or between an end frame and the wall of a hive. Brace Comb is a form of burr comb.

BROOD: A general term to describe young bees that have not yet emerged from their cells as adults. Brood can be workers, drones or queens. The four stages of brood are egg, larva, pre-pupa and pupa.

BROOD CHAMBER: The area of the hive where the queen lays eggs and brood is reared, usually the lowermost parts of the hive.

CAPPED BROOD: Brood in the pre-pupal or pupal stage inside a capped cell. Immature bees whose cells have been capped with a brown wax cover by other worker bees; inside the sealed cell, the non-feeding larvae are isolated and can spin cocoons before pupating.

CAPPED HONEY: Cells full of ripe or mature honey, sealed or capped with beeswax.

CAPPED LARVA: The stage of brood development when the larvae turn into pre-pupae and are ready to have its cell capped and to spin its silk cocoon, about the tenth day from the laying of the egg.

CASTES: The three types of bees that comprise the adult population of a honey bee colony are the queen, workers, and drones.

CHRONIC PARALYSIS VIRUS (CPV): Sometimes called chronic bee paralysis virus, this is a disease mainly of adult bees.

CELL CUP: The base of an artificial queen cell made of beeswax or plastic and used for rearing queen bees.

CLEARER BOARD: A board used to clear bees from a super before removing the frames ready for extraction.

CLIPPED QUEEN: Queen whose wing (or wings) has been clipped to stop her from swarming.

COLONY: A collection or family of bees living within a single social unit containing workers, drones and a queen.

COMB FOUNDATION: A thin sheet of beeswax with the base pattern of cells impressed on the sheet as a template for the bees to start making comb. Some foundation is also made of plastic.

COMB HONEY: Honey is sold and eaten in the comb and also referred to as cut comb honey.

DEFORMED WING VIRUS (DWV): A virus of the honey bee, often associated with a *Varroa* infestation, that causes newly emerged bees to have deformed wings.

DISEASE RESISTANCE: The ability of an organism to avoid a particular disease, primarily due to genetic immunity or avoidance behaviour.

DOUBLE: A hive that consists of two boxes.

DRIFT: The process in which bees join a different hive from their own, often due to loss of direction or when hives are placed too close together.

DRONE CONGREGATION AREA (DCA): An area where many drones from surrounding colonies gather to mate with queens during their nuptial or mating flights. The location of drone congregation areas has remained stable over many years.

DYSENTERY: An abnormal condition of adult bees characterized by severe diarrhoea and usually caused by starvation, low-quality food, moist surroundings or *Nosema apis* infection. (See also *Nosema*)

EMERGENCY QUEEN CELL: A queen cell in the middle of a brood frame that has been made quickly due to the queen bee dying suddenly.

ENTRANCE REDUCER: A device used to regulate the size of the bottom entrance. May be used to reduce hive entrance during winter months or attacks from robber bees or wasps.

ESCAPE BOARD: A board with one or more one-way bees escapes. Used to empty a super of bees.

ESCORTS OR ESCORT BEES: Worker bees that are placed in a cage with a queen for shipping. Usually about four or five escort bees are included with the queen.

EUROPEAN FOUL BROOD (EFB): An infectious disease of honey bee brood caused by the bacterium *Streptococcus pluton.*

EXTRACTION: Removal of honey from comb. Typically refers to the use of an extractor but also includes non-mechanical methods such as crush and strain.

FAILING QUEEN: A queen that has reached the end of her fertile age and is not producing sufficient brood for the colony to survive. The colony is now ready for the supersedure of the queen.

FERAL BEES: Bees that are not kept in hives by beekeepers.

FERTILE QUEEN: A queen, inseminated instrumentally or mated with a drone, which can lay fertilized eggs.

GUARDING: Bees stationed at the entrance of a hive to detect and ward off invaders and to examine entering bees. Guard bees are typically about three weeks old.

HIVE STAND: A device that raises the bottom board off the ground and helps extend the life of the bottom board by keeping it dry.

HONEY BEE RACE: Honey bee race is a classification of honey bees, in particular the European honey bee, *Apis mellifera*, into different types. The most common races of *Apis mellifera* are Italian bees, Caucasian bees, Carniolan bees and dark German bees.

HONEY BOUND: A brood nest that is congested by cells/comb filled with honey, reducing the amount of space that the queen has to lay eggs.

HONEY FLOW: A time, usually in the spring and summer, when there are sufficient nectar-bearing plants blooming so that bees can store a surplus of honey.

HONEY SUPERS: The hive bodies or boxes, usually above the brood box, used by the workers to store honey.

HONEYDEW: The sweet secretion from aphids and scale insects. Since honeydew contains almost 90 per cent sugar; it is collected by bees and stored as honeydew honey.

HOOP PINE: A high quality pine wood native to New South Wales and Queensland and used to make supers and other hive parts.

HOPELESSLY QUEENLESS: A colony of bees that has been without a queen for several weeks and, as a result, is unable to make a new queen from young larvae.

HYDROGEN PEROXIDE: Found in all honeys and provides most of its antibacterial properties.

INBREEDING: Mating among related individuals that may cause genetic deformities in the bee.

INFERTILE: A bee that is unable to produce a fertilized egg, typically applied to a laying worker.

INTEGRATED PEST MANAGEMENT (IPM): A pest control method that uses a variety of complementary strategies including genetic, biological, cultural management, chemical management, as well as mechanical and physical devices such as screened bottom boards. IPM techniques are typically performed in three stages: prevention, observation and intervention. It is an ecologically-friendly approach with the goal of significantly reducing or eliminating the use of pesticides while at the same time managing pest populations at an acceptable level.

INTERLOCKING: A design of wooden hive in which the corners of the boxes interlock with each other.

INTRODUCING: The process of introducing a new queen into an existing hive.

INTRODUCING CAGE: A small wood, wire or plastic cage used to ship queens and also sometimes to release them into the colony. Also known as a queen cage.

IRRADIATE: To irradiate a hive with gamma radiation to kill American Foul Brood or other diseases.

ISLE OF WIGHT DISEASE: *Acarapis woodi*, a small mite that causes tracheal infections of the honey bee, is believed to have been the cause of the severe decline in the honey bee population on the Isle of Wight, off the south coast of England, in 1904.

ISRAELI ACUTE PARALYSIS VIRUS (IAPV): First identified in Israel but now infecting honey bees in many parts of the world, one of several viruses that causes paralysis of adult honey bees.

ITALIAN BEE: A common race of bees, *Apis mellifera ligustica*, which originated in Italy. Bees have brown and yellow bands; usually gentle and productive, but tend to rob other hives. Italian bees are the most well known type of bee.

KASHMIR BEE VIRUS (KBV): A naturally occurring virus infecting *Apis cerana*, Kashmir Bee Virus jumped species to infect *Apis mellifera* and is now a common infection of this bee.

KILLER BEES: Africanized bee.

LARVA: The second stage of development in the lifecycle of the bee. The three stages are egg, larva and pupa. The larva stage is often called a grub.

LARVAE: The plural of larva.

LIFTING CLEAT: A wooden cleat attached to the side of supers and brood boxes to make them easier to lift.

LINE: A family or set of descendants bred from the same queen.

MANIPULATIONS: The process by which frames in a hive are moved or replaced.

MATED QUEEN: A queen that has gone on a mating flight or has been artificially inseminated with drone sperm.

MATING FLIGHT: A short flight taken by a virgin queen during which she mates in the air with several drones from other colonies. Queens usually mate with ten to twenty drones on one or more mating flights.

MIGRATORY BEEKEEPING: The practice of professional beekeepers and some hobby beekeepers to regularly move their hives to follow the flowering pattern of different types of trees.

MIGRATORY LID: The lid of a hive that has the same outside dimensions as the super. Called a migratory lid since it is of a convenient size to allow beekeepers to transport their hives on the back of trucks. A lid that does not extend over the sides of the hive.

MITICIDE: A chemical or biological agent which is applied to a colony to control parasitic mites such as *Varroa* or Tracheal mite.

NATIVE BEES: Bees that are native to a country, for example, stingless bees. The European honey bee is not a native bee in most of the world as it was introduced by European settlers.

NATIVE POLLINATORS: Apart from native bees, other animals may also be native pollinators such as birds or other insects. (See Native Bees)

NATURAL BEEKEEPING: A philosophy of keeping bees that puts the welfare of the colony ahead of the amount of honey collected during a season.

NATURAL HONEY: Pure honey that has not been heated or finely filtered during processing prior to bottling.

NECTAR FLOW: The time of the year when a tree or shrub produces a lot of nectar for bees to collect. The nectar flow of a particular tree often only lasts a few weeks, so professional beekeepers need to know what trees are flowering where and when and move their hives to follow the flowering trees.

NEONICOTINOIDES: A class of insecticides chemically related to nicotine.

NEST: A feral colony of bees that is not being managed by a beekeeper.

NEWSPAPER METHOD: A technique to join two unrelated colonies by providing a temporary newspaper barrier. The bees will chew their way through the newspaper over a few days and the two colonies should merge without fighting.

NOSEMA: A disease caused by protozoan spore-forming parasites living in the gut of adult bees. The two types of nosema that infect adult bees are *Nosema apis* and *Nosema cerana.* (See *Nosemosis*)

NOSEMOSIS: To be infected with *Nosema.*

NUC or NUCLEUS HIVE: A colony of bees housed within a small brood box usually containing only four or five frames. Nucleus hives are often used to rear queens.

NURSERY BEES: Young bees, three to ten days old, which feed and take care of developing larvae.

OBSERVATION HIVE: A hive with walls mainly made of glass or clear plastic to allow the observation of bees at work and often used as a teaching aid during courses.

ORGANIC HONEY: Honey made from nectar from flowers grown in organically certified regions.

ORIENTATION FLIGHTS: Short orienting flights taken by young bees, usually by large numbers at the same time and during the warm part of day, in order for them to learn to fly and to familiarize themselves with their surroundings.

OVERCROWDING: A hive that has too many bees living in it for its size.

OVERWINTERING: The survival of a colony over winter.

OXYTETRACYCLINE (OTC): An antibiotic sold under the trade name Terramycin. Oxytetracycline is used to control American and European Foul Brood diseases.

PARASITIC MITE SYNDROME (PMS): Signs of poor health in a colony of bees that has been infested with *Varroa.*

PARENT COLONY: The home colony from which swarms or splits originated.

PHEROMONE: Several kinds of scents produced by bees to establish a basic form of communication or to stimulate a response.

PIPING: A series of sounds made by a queen, usually before she emerges from her cell.

PLANTATION PINE: The most readily available plantation timbers are softwoods such as hoop pine, radiata pine and slash pine.

PLASTIC FOUNDATION: Foundation placed inside a frame made entirely of plastic, used as an alternative to wax foundation. (See Plastic Frames)

PLASTIC FRAMES: Frames that are constructed entirely of plastic, including the outer rim.

PLAY FLIGHTS: Short flights taken in front and in the vicinity of the hive by young bees to acquaint themselves with the location of the hive. Play flights are sometimes mistaken for robbing or preparation for swarming. (See Orientation Flights)

POLLEN: The male reproductive cells of flowers. Pollen provides the protein in a young bee's diet and is frequently called bee bread when stored in cells in the colony. Pollen is an essential component of brood food. Honey is another essential component and provides the carbohydrate part of the bee's diet.

POLLEN BASKET: The area on the hind leg of bee adapted to carrying pellets of pollen or propolis back to the colony.

POLLEN SUBSTITUTE: A food material which is used as a substitute for naturally occurring pollen. Pollen substitute usually contains soy flour, brewers' yeast, powdered sugar and other ingredients. Pollen substitute is used to stimulate brood rearing in periods of pollen shortage.

POLLEN TRAP: Device which forces the bees entering a hive to walk through a meshed screen; the pollen is brushed off the bees' legs by the screen and is collected from a collecting tray.

PRIME SWARM: The first swarm to leave the parent colony, usually with the old queen. (See also After-swarm)

PROPOLIS: A sticky resinous material collected from trees or other plants by bees; used to close holes and cover surfaces in the hive. Often referred to as bee glue. Propolis also has antimicrobial properties and is used by bees for this purpose.

PROPOLISE: To fill with propolis, or bee glue; used by bees to strengthen comb and seal cracks.

PROTEIN: Naturally occurring complex organic substances, such as pollen, and composed of amino acids. An essential food for brood to build body tissue before emerging as adult bees.

PUPA: The final stage in the life of a developing baby bee after larva and before maturity. Pupae are only found inside sealed brood comb. Pupae is the plural of pupa.

PYRETHRINS: A group of naturally occurring insecticides derived from the seed cases of the perennial plant *Chrysanthemum cinerariaefolium.*

QUEENLESS COLONY: A colony that does not have a queen.

QUEEN-RIGHT: A colony of bees that contains a laying queen.

RE-QUEEN: To introduce a new queen to a queenless hive. Usually to replace an old queen with a young one.

ROBBER BEES: Bees that rob other colonies of their honey. (See Robbing)

ROBBING: Bees that are stealing honey from other hives. This is a common problem particularly in autumn when nectar is not available in the field. The term also applies to bees that are cleaning out wet supers or cappings left uncovered by beekeepers.

ROPEY CHARACTERISTIC: A diagnostic test for American Foul Brood and European Foul Brood in which the decayed larvae or pupae form an elastic rope when drawn out with a matchstick.

ROYAL JELLY: Bee milk and worker jelly all refer to royal jelly, which is a pearly white, creamy substance produced by young worker bees to feed larvae. Royal jelly is a secretion of the hypopharyngeal glands located in the head of young worker honey bees.

SACBROOD: A viral disease of larvae, usually nonfatal to the colony, which interferes with the moulting process; the dead larva resembles a bag of fluid.

SCOUT BEES: Worker bees out searching for a new source of pollen, nectar, propolis, water or a new home for a swarm of bees.

SCREENED BOTTOM BOARD: A framed screen used instead of a solid base in order to improve ventilation through the hive. Also used as a means to control *Varroa*, SHB and to allow debris to fall through to the outside of the hive.

SEALED BROOD: See Capped Brood.

SECTIONS: Small plastic or wooden boxes placed inside frames and used to produce comb honey.

SKEP HIVE: Traditional hive typically made of straw resembling an upturned basket in which bees build comb. Illegal in many countries because comb can't be inspected for diseases.

SLUM-GUM: The black, sticky, fibrous, oily or waxy material left when comb has been melted and filtered by a beekeeper.

SMALL HIVE BEETLE (SHB): The Small Hive Beetle *(Aethina tumida)* is a destructive pest of honey bee colonies, causing damage to comb, stored honey and pollen. If infestation is sufficiently heavy, this may cause the bees to abandon their hive. The beetles can also be a pest of stored combs, and honey (in the comb) awaiting extraction. Beetle larvae tunnel through combs of honey, feeding and defecating, causing discoloration and fermentation.

SNOTTY BROOD: Sick brood with physical symptoms very similar to European Foul Brood (EFB). Snotty brood is also called 'Idiopathic Brood Disease Syndrome', which is a fancy way of saying that we haven't figured out what causes it yet and we are still looking.

SOLAR WAX MELTER: A glass-covered box in which wax combs and cappings are melted by the sun's rays and wax is recovered in cake form.

SPOTTED BROOD: An irregular brood pattern on the frame caused by disease or a failing queen.

SPLIT: To divide a colony in two to increase the number of hives.

SPRING BUILD-UP: The increase in the number of bees in a hive during very late winter and spring.

SPRING DWINDLING: A decrease in the size of the colony population during spring instead of the significant population growth which should be experienced. Spring dwindling is not associated with swarming in which about half the colony leave the hive to found a new colony.

STICKY: A super frame that has had its honey removed and is now covered in honey and is sticky to touch.

STREPTOCOCCUS PLUTON: Bacterium that causes European Foul Brood (EFB).

SUPER: A wooden box with frames containing foundation or drawn comb in which honey is to be produced. Named for its position above the brood nest, i.e., superstructure. The same size box is referred to as a hive body or brood box when it is situated below the honey supers and is to be used for brood rearing.

SUPERING: Placing additional supers on a hive in order to collect excess honey and provide more room for workers in a crowded hive.

SUPERSEDURE: The replacement of a weak or old queen in a colony by a daughter queen. Shortly after the daughter queen begins to lay eggs, the mother queen disappears or is killed by the other bees.

SWARM: A collection of bees containing at least one queen that has left its home colony to look for a new nest. Swarms are currently without a home site and are looking for a new one, often resting on a tree or other object.

SWARM CELL: Queen cells usually found on the bottom of the combs before swarming.

SWARMING: The natural method by which bee colonies propagate. When the hive or nest becomes congested with bees in the spring, about half the colony flies off with the old queen to find a new home, leaving a new queen behind to head the old colony.

SWARMING SEASON: The time of year, usually between early spring and mid-summer, when colonies swarm.

SYRUP: A mixture of sugar and water used to feed bees when nectar sources are scarce or the beekeeper wants the colony to make honey for the winter.

TESTED QUEEN: A queen whose offspring have been tested to show that she has mated with a drone of her own race or has other qualities which would make her a good colony mother.

TOP BAR HIVE (TBH): The top bar hive is a method to manage bees with removable combs which rely on top bars rather than frames for the combs. There are no standard dimensions for a TBH as there are for Langstroth hives. TBHs have some advantages for hobby beekeepers. (Also known as Kenyan top bar hive and Tanzanian top bar hive.)

TOP FEEDER: A syrup feeder that is placed at the top of the hive under the lid.

TRACHEAL MITE: A small mite, *Acarapis woodi*, that infests the tracheas of the honey bee.

TRIPLE: A hive consisting of three boxes. This configuration may be one brood box and two supers, or a single brood box and two supers.

UNDERSUPERING: Placing an additional super beneath an existing super but above the brood chamber.

UNITE: Combining two or more colonies to form a larger colony.

UNRIPE: Honey in cells that have not been capped because the workers have not yet evaporated sufficient water to turn the nectar into honey. If unripe honey is extracted it can ferment.

UNSEALED BROOD: Brood during the egg and larval stages that are not yet in sealed cells.

UNSEALED HONEY: Honey in cells that have not yet been capped. (See also Unripe).

VARROA DESTRUCTOR: An external parasitic mite that attacks *Apis cerana* and *Apis mellifera* and develops in sealed brood cells. *Varroa destructor* is closely related to *Varroa jacobsoni.*

VARROA JACOBSONI: An external parasitic mite that attacks *Apis cerana* and *Apis mellifera* and develops in sealed brood cells. *Varroa jacobsoni* is closely related to *Varroa destructor.*

***VARROA* SENSITIVE HYGIENE (VSH)**: A genetically selected honey bee which appears to suppress *Varroa* mite reproduction.

VENT: A perforated piece of metal used to cover the air vent in the lid of a hive to stop robber bees or wasps entering.

VENTILATED LID: A hive lid that has vents in the rim allowing air to circulate.

VIRGIN QUEEN: A queen that has not yet left the colony to go on a mating flight and is thus not yet able to produce offspring.

WASHBOARDING: Worker honey bees exhibit a group activity known as rocking or washboarding on the internal and external surfaces of the hive. This behaviour is believed to be associated with general cleaning activities but under what circumstances workers washboard is not known.

WAX DIPPED: Hive parts that have been dipped in hot wax to protect the wood from the weather. Frames are not typically dipped in wax to protect them.

WAX GLANDS: Eight glands located on the underside of a bee's abdomen from which wax is secreted.

WAX MOTH: Usually refers to the Greater Wax Moth (GWM), *Galleria mellonella*, whose larvae bore through and destroy honeycomb as they eat out its impurities. Wax Moth infested hives are covered in a thick, spider-web like material. The Lesser Wax Moth (LWM), *Achroia grisella*, is the smaller of the two related species. Its larvae are not found in the large congregated numbers of the greater wax moth and are often solitary.

WEATHERTEX: A range of long lasting processed wood fibres that have been rolled into sheets. Weathertex is frequently used to make hive bases and hive lids.

WINTER CLUSTER: A closely packed colony of bees within a hive that forms a cluster to conserve heat when outside temperature falls below about 15° Celsius.

WINTER HARDINESS: The ability of some strains of honey bees to survive long winters by minimizing their use of stored honey; the northern European black honey bee would be an example of this.

WINTERING DOWN: Preparing a hive for the winter.

WIRED FRAMES: Frames with wires holding sheets of foundation in place.

WORKER: An unfertilized female bee that makes up the majority of a colony's population.

WORKER JELLY: See Royal Jelly.

WSP: A super or brood box with a depth of 192mm.

INDEX

Page numbers in **bold** refer to images.

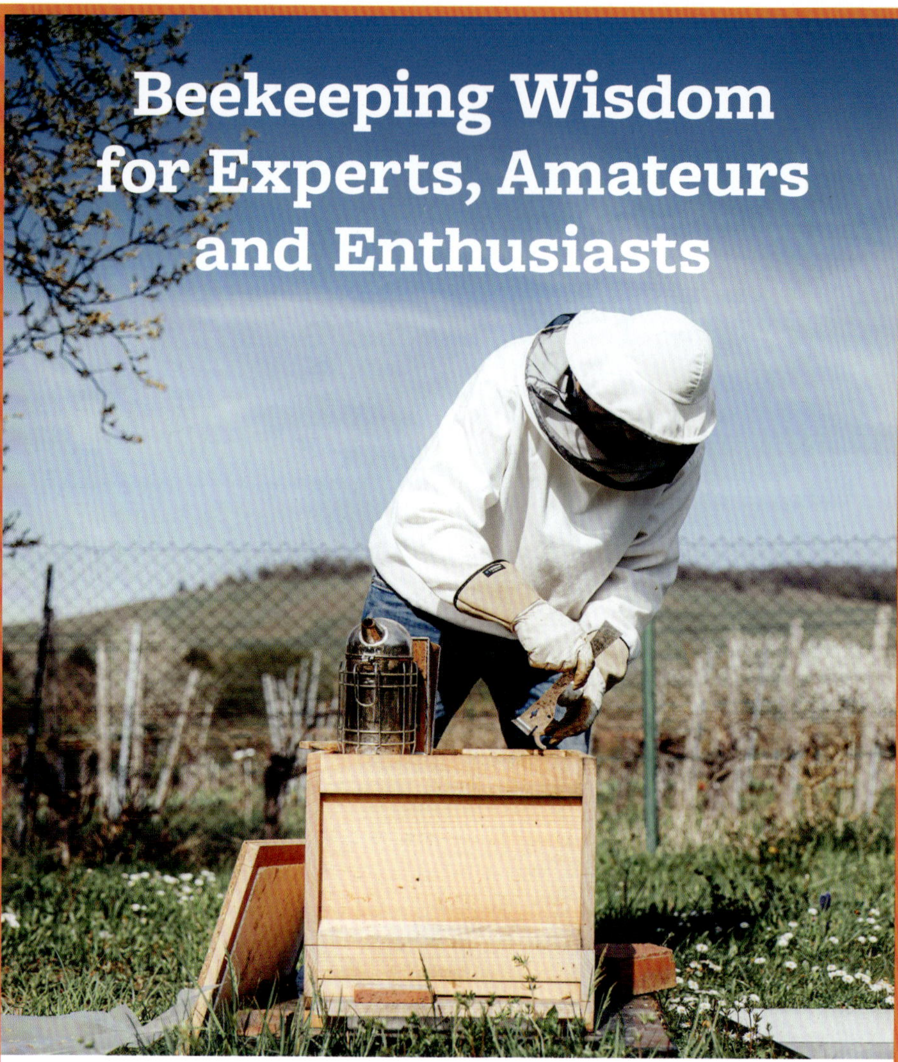